AN ANALYSIS OF
THE LEVER ESCAPEMENT

AN ANALYSIS OF THE LEVER ESCAPEMENT

H. R. PLAYTNER

ISBN: 978-93-5420-051-9

Published: 1910

LECTOR HOUSE LLP
E-MAIL: lectorpublishing@gmail.com

THOMAS MUDGE

*The first Horologist who successfully applied the Detached
Lever Escapement to Watches.
Born 1715 — Died 1794.*

AN ANALYSIS OF
THE LEVER ESCAPEMENT

A LECTURE DELIVERED BEFORE
THE CANADIAN WATCHMAKERS'
AND RETAIL JEWELERS' ASSOCIATION.

BY

H. R. PLAYTNER.

ILLUSTRATED.

1910.

PREFACE.

Before entering upon our subject proper, we think it advisable to explain a few points, simple though they are, which might cause confusion to some readers. Our experience has shown us that as soon as we use the words "millimeter" and "degree," perplexity is the result. "What is a millimeter?" is propounded to us very often in the course of a year; nearly every new acquaintance is interested in having the metric system of measurement, together with the fine gauges used, explained to him.

The metric system of measurement originated at the time of the French Revolution, in the latter part of the 18th century; its divisions are decimal, just the same as the system of currency we use in this country.

A meter is the ten millionth part of an arc of the meridian of Paris, drawn from the equator to the north pole; as compared with the English inch there are $39^{3708}/_{10000}$ inches in a meter, and there are 25.4 millimeters in an inch.

The meter is sub-divided into decimeters, centimeters and millimeters; 1,000 millimeters equal one meter; the millimeter is again divided into 10ths and the 10ths into 100ths of a millimeter, which could be continued indefinitely. The $\frac{1}{100}$ millimeter is equal to the $\frac{1}{2540}$ of an inch. These are measurements with which the watchmaker is concerned. $\frac{1}{100}$ millimeter, written .01 mm., is the side shake for a balance pivot; multiply it by 2¼ and we obtain the thickness for the spring detent of a pocket chronometer, which is about ⅓ the thickness of a human hair.

The metric system of measurement is used in all the watch factories of Switzerland, France, Germany, and the United States, and nearly all the lathe makers number their chucks by it, and some of them cut the leading screws on their slide rests to it.

In any modern work on horology of value, the metric system is used. Skilled horologists use it on account of its *convenience*. The millimeter is a unit which can be handled on the small parts of a watch, whereas the inch must always be divided on anything smaller than the plates.

Equally as fine gauges can be and are made for the inch as for the metric system, and the inch is decimally divided, but we require another decimal point to express our measurement.

Metric gauges can now be procured from the material shops; they consist of tenth measures, verniers and micrometers; the finer ones of these come from Glashutte, and are the ones mentioned by Grossmann in his essay on the lever escapement. Any workman who has once used these instruments could not be persuaded to do without them.

No one can comprehend the geometrical principles employed in escapements without a knowledge of angles and their measurements, therefore we deem it of sufficient importance to at least explain what a degree is, as we know for a fact, that young workmen especially, often fail to see how to apply it.

Every circle, no matter how large or small it may be, contains 360°; a degree is therefore the 360th part of a circle; it is divided into minutes, seconds, thirds, etc.

To measure the *value* of a degree of any circle, we must multiply the diameter of it by 3.1416, which gives us the circumference, and then divide it by 360. It will be seen that it depends on the size of that circle or its radius, as to the value of a degree in any *actual* measurement. To illustrate; a degree on the earth's circumference measures 60 geographical miles, while measured on the circumference of an escape wheel 7.5 mm. in diameter, or as they would designate it in a material shop, No. 7½, it would be 7.5 × 3.1416 ÷ 360 = .0655 mm., which is equal to the breadth of an ordinary human hair; it is a degree in both cases, but the difference is very great, therefore a degree cannot be associated with any actual measurement until the radius of the circle is known. Degrees are generated from the center of the circle, and should be thought of as to ascension or direction and relative value. Circles contain four right angles of 90° each. Degrees are commonly measured by means of the protractor, although the ordinary instruments of this kind leave very much to be desired. The lines can be verified by means of the compass, which is a good practical method.

It may also be well to give an explanation of some of the terms used.

Drop equals the amount of freedom which is allowed for the action of pallets and wheel. See Z, Fig. 1.

Primitive or Geometrical Diameter.—In the ratchet tooth or English wheel, the primitive and real diameter are equal; in the club tooth wheel it means across the locking corners of the teeth; in such a wheel, therefore, the primitive is *less* than the real diameter by the height of two impulse planes.

Lock equals the depth of locking, measured from the locking corner of the pallet at the moment the drop has occurred.

Run equals the amount of angular motion of pallets and fork to the bankings *after* the drop has taken place.

Total Lock equals lock plus run.

A *Tangent* is a line which *touches* a curve, but does not intersect it. AC and AD, Figs. 2 and 3, are tangents to the primitive circle GH at the points of intersection of EB, AC, and GH and FB, AD and GH.

Impulse Angle equals the angular connection of the impulse or ruby pin with the lever fork; or in other words, of the balance with the escapement.

Impulse Radius.—From the face of the impulse jewel to the center of motion, which is in the balance staff, most writers assume the impulse angle and radius to be equal, and it is true that they must conform with one another. We have made a radical change in the radius and one which does not affect the angle. We shall prove this in due time, and also that the wider the impulse pin the greater must the impulse radius be, although the angle will remain unchanged.

Right here we wish to put in a word of advice to all young men, and that is to learn to draw. No one can be a thorough watchmaker unless he can draw, because he cannot comprehend his trade unless he can do so.

We know what it has done for us, and we have noticed the same results with others, therefore we speak from personal experience. Attend night schools and mechanic's institutes and improve yourselves.

The young workmen of Toronto have a great advantage in the Toronto Technical School, but we are sorry to see that out of some 600 students, only five watchmakers attended last year. We can account for the majority of them, so it would seem as if the young men of the trade were not much interested, or thought they could not apply the knowledge to be gained there. This is a great mistake; we might almost say that knowledge of any kind can be applied to horology. The young men who take up these studies, will see the great advantage of them later on; one workman will labor intelligently and the other do blind "guess" work.

We are now about to enter upon our subject and deem it well to say, we have endeavored to make it as plain as possible. It is a deep subject and is difficult to treat lightly; we will treat it in our own way, paying special attention to all these points which bothered us during the many years of painstaking study which we gave to the subject. We especially endeavor to point out how theory can be applied to practice; while we cannot expect that everyone will understand the subject without study, we think we have made it comparatively easy of comprehension.

We will give our method of drafting the escapement, which happens in some respects to differ from others. We believe in making a drawing which we can reproduce in a watch.

CONTENTS

AN ANALYSIS
OF THE LEVER ESCAPEMENT.

LEVER ESCAPEMENT.

The lever escapement is derived from Graham's dead-beat escapement for clocks. Thomas Mudge was the first horologist who successfully applied it to watches in the detached form, about 1750. The locking faces of the pallets were arcs of circles struck from the pallet centers. Many improvements were made upon it until to-day it is the best form of escapement for a general purpose watch, and when made on mechanical principles is capable of producing first rate results.

Our object will be to explain the whys and wherefores of this escapement, and we will at once begin with the number of teeth in the escape wheel. It is not obligatory in the lever, as in the verge, to have an uneven number of teeth in the wheel. While nearly all have 15 teeth, we might make them of 14 or 16; occasionally we find some in complicated watches of 12 teeth, and in old English watches, of 30, which is a clumsy arrangement, and if the pallets embrace only three teeth in the latter, the pallet center cannot be pitched on a tangent.

Although advisable from a timing standpoint that the teeth in the escape wheel should divide evenly into the number of beats made per minute in a watch with seconds hand, it is not, strictly speaking, necessary that it should do so, as an example will show. We will take an ordinary watch, beating 300 times per minute; we will fit an escape wheel of 16 teeth; multiply this by 2, as there is a forward and then a return motion of the balance and consequently two beats for each tooth, making $16 \times 2 = 32$ beats for each revolution of the escape wheel. 300 beats are made per minute; divide this by the beats made on each revolution, and we have the number of times in which the escape wheel revolves per minute, namely, $300 \div 32 = 9.375$. This number then is the proportion existing for the teeth and pitch diameters of the 4th wheel and escape pinion. We must now find a suitable number of teeth for this wheel and pinion. Of available pinions for a watch, the only one which would answer would be one of 8 leaves, as any other number would give a fractional number of teeth for the 4th wheel, therefore $9.375 \times 8 = 75$ teeth in 4th wheel. Now as to the proof: as is well known, if we multiply the number of teeth contained in 4th and escape wheels also by 2, for the reason previously given, and divide by the leaves in the escape pinion, we get the number of beats made per minute; therefore $\frac{(75 \times 16 \times 2)}{8} = 300$ beats per minute.

Pallets can be made to embrace more than three teeth, but would be much heavier and therefore the mechanical action would suffer. They can also be made to embrace fewer teeth, but the necessary side shake in the pivot holes would prove very detrimental to a total lifting angle of 10°, which represents the angle of movement in modern watches. Some of the finest ones only make 8 or 9° of a movement; the smaller the angle the greater will the effects of defective workmanship be; 10° is a common-sense angle and gives a safe escapement capable of fine results. Theoretically, if a timepiece could be produced in which the balance would vibrate without being connected with an escapement, we would have reached a step nearer the goal. Practice has shown this to be the proper theory to work on. Hence, the smaller the pallet and impulse angles the less will the balance and escapement be connected. The chronometer is still more highly detached than the lever.

The pallet embracing three teeth is sound and practical, and when applied to a 15 tooth wheel, this arrangement offers certain geometrical and mechanical advantages in its construction, which we will notice in due time. 15 teeth divide evenly into 360° leaving an interval of 24° from tooth to tooth, which is also the angle at which the locking faces of the teeth are inclined from the center, which fact will be found convenient when we come to cut our wheel.

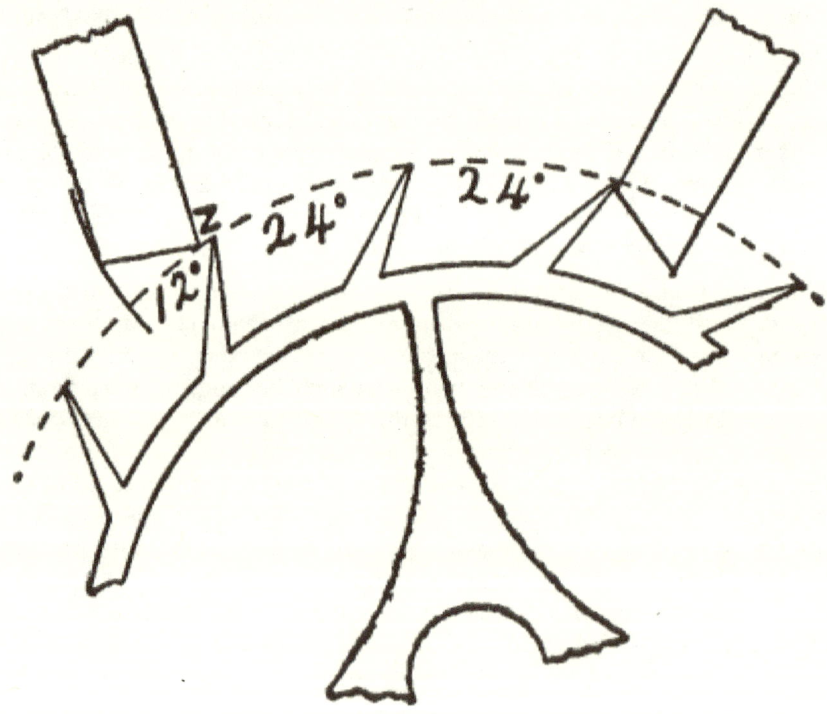

FIG. 1.

From locking to locking on the pallet scaping over three teeth, the angle is 60°, which is equal to 2½ spaces of the wheel. Fig. 1 illustrates the lockings, spanning this arc. If the pallets embraced 4 teeth, the angle would be 84°; or in case of a 16 tooth wheel scaping over three teeth, the angle would be $360 \times {}^{2.5}\!/_{16} = 56\frac{1}{4}°$.

Pallets may be divided into two kinds, namely: equidistant and circular. The equidistant pallet is so-called because the lockings are an equal distance from the center; sometimes it is also called the tangential escapement, on account of the unlocking taking place on the intersection of tangent AC with EB, and FB with AD, the tangents, which is the valuable feature of this form of escapement.

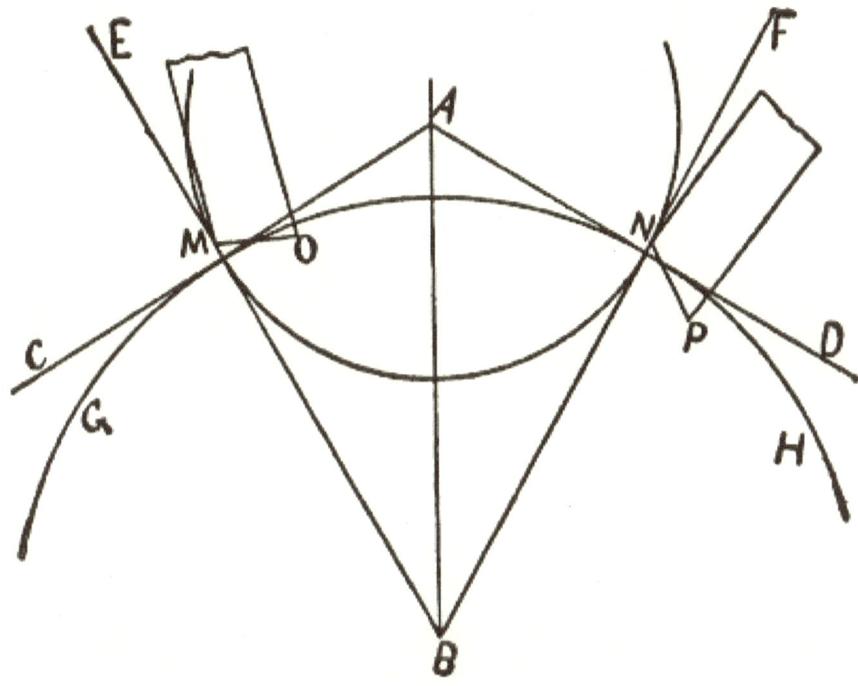

FIG. 2.

AC and AD, Fig. 2, are tangents to the primitive circle GH. ABE and ABF are angles of 30° each, together therefore forming the angle FBE of 60°. The locking circle MN is struck from the pallet center A; the interangles being equal, consequently the pallets must be equidistant.

The weak point of this pallet is that the lifting is not performed so favorably; by examining the lifting planes MO and NP, we see that the discharging edge, O, is closer to the center, A, than the discharging edge, P; consequently the lifting on the engaging pallet is performed on a shorter lever arm than on the disengaging pallet, also any inequality in workmanship would prove more detrimental on the engaging than on the disengaging pallet. The equidistant pallet requires fine workmanship throughout. We have purposely shown it of a width of 10°, which is the widest we can employ in a 15 tooth wheel, and shows the defects of this escapement more readily than if we had used a narrow pallet. A narrower pallet is advisable, as the difference in the discharging edges will be less, and the lifting arms would, therefore, not show so much difference in leverage.

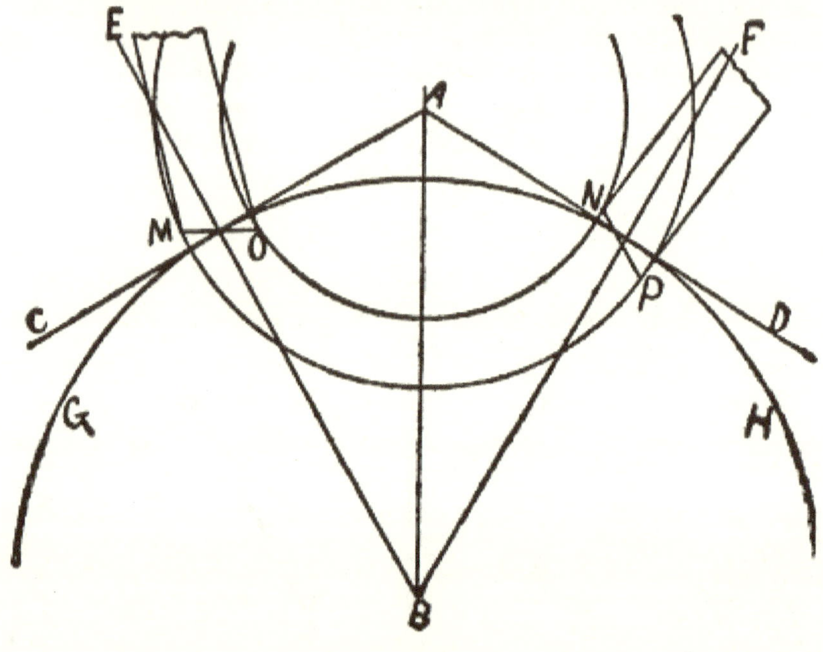

FIG. 3.

The circular pallet is sometimes appropriately called "the pallet with equal lifts," as the lever arms AMO and ANP, Fig. 3, are equal lengths. It will be noticed by examining the diagram, that the pallets are bisected by the 30° lines EB and FB, one-half their width being placed on each side of these lines. In this pallet we have two locking circles, MP for the engaging pallet, and NO for the disengaging pallet. The weak points in this escapement are that the unlocking resistance is greater on the engaging than on the disengaging pallet, and that neither of them lock on the tangents AC and AD, at the points of intersection with EB and FB. The narrower the circular pallet is made, the nearer to the tangent will the unlocking be performed. In neither the equidistant or circular pallets can the unlocking resistance be *exactly* the same on each pallet, as in the engaging pallet the friction takes place before AB, the line of centers, which is more severe than when this line has been passed, as is the case with the disengaging pallet; this fact proportionately increases the existing defects of the circular over the equidistant pallet, and *vice versa*, but for the same reason, the lifting in the equidistant is proportionately accompanied by more friction than in the circular.

Both equidistant and circular pallets have their adherents; the finest Swiss, French and German watches are made with equidistant escapements, while the majority of English and American watches contain the circular. In our opinion the English are wise in adhering to the circular form. We think a ratchet wheel should not be employed with equidistant pallets. By examining Fig. 2, we see an English pallet of this form. We have shown its defects in such a wide pallet as the English (as we have before stated), because they are more readily perceived; also, on account of the shape of the teeth, there is danger of the discharging edge, P, dipping so deep into the wheel, as to make considerable drop necessary, or the pallets would touch on the backs of the teeth. In the case of the club tooth, the latter is hollowed out, therefore, less drop is required. We have noticed that theoretically, it is advantageous to make the pallets narrower than the English, both for the equidistant and circular escapements. There is an escapement, Fig. 4, which is just the opposite to the English. The entire lift is performed by the wheel, while in the case of the ratchet wheel, the entire lifting angle is on the pallets; also, the pallets being as narrow as they can be made, consistent with strength, it has the good points of both the equidistant and circular pallets, as the unlocking can be performed on the tangent and the lifting arms are of equal length. The wheel, however, is so much heavier as to considerably increase the inertia; also, we have a metal surface of quite an extent sliding over a thin jewel. For practical reasons, therefore, it has been slightly altered in form and is only used in cheap work, being easily made.

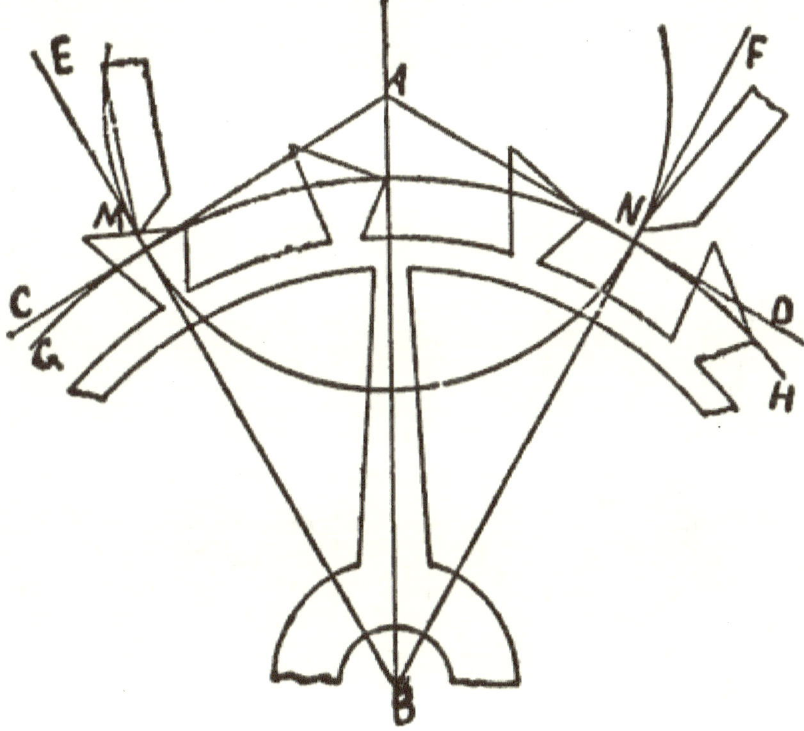

FIG. 4.

We will now consider the drop, which is a clear loss of power, and, if excessive, is the cause of much irregularity. It should be as small as possible consistent with perfect freedom of action.

In so far as *angular* measurements are concerned, no hard and fast rule can be applied to it, the larger the escape wheel the smaller should be the angle allowed for drop. Authorities on the subject allow 1½° drop for the club and 2° for the ratchet tooth. It is a fact that escape wheels are not cut perfectly true; the teeth are apt to bend slightly from the action of the cutters. The truest wheel can be made of steel, as each tooth can be successively ground after being hardened and tempered. Such a wheel would require less drop than one of any other metal. Supposing we have a wheel with a primitive diameter of 7.5 mm., what is the amount of drop, allowing 1½° by angular measurement? 7.5 × 3.1416 ÷ 360 × 1.5 = .0983 mm., which is sufficient; a hair could get between the pallet and tooth, and would not stop the watch. Even after allowing for imperfectly divided teeth, we require no greater freedom even if the wheel is larger. Now suppose we take a wheel with a primitive diameter of 8.5 mm. and find the amount of drop; 8.5 × 3.1416 ÷ 360 × 1.5 = .1413 mm., or .1413 − .0983 = .043 mm., more drop than the smaller wheel, if we take the same angle. This is a waste of force. The angular drop should, therefore, be proportioned according to the size of the wheel. We wish it to be understood that common sense must always be our guide. When the horological student once arrives at this standpoint, he can *intelligently* apply himself to his calling.

THE DRAW.

The Draw. — The draw or draft angle was added to the pallets in order to draw the fork back against the bankings and the guard point from the roller whenever the safety action had performed its function.

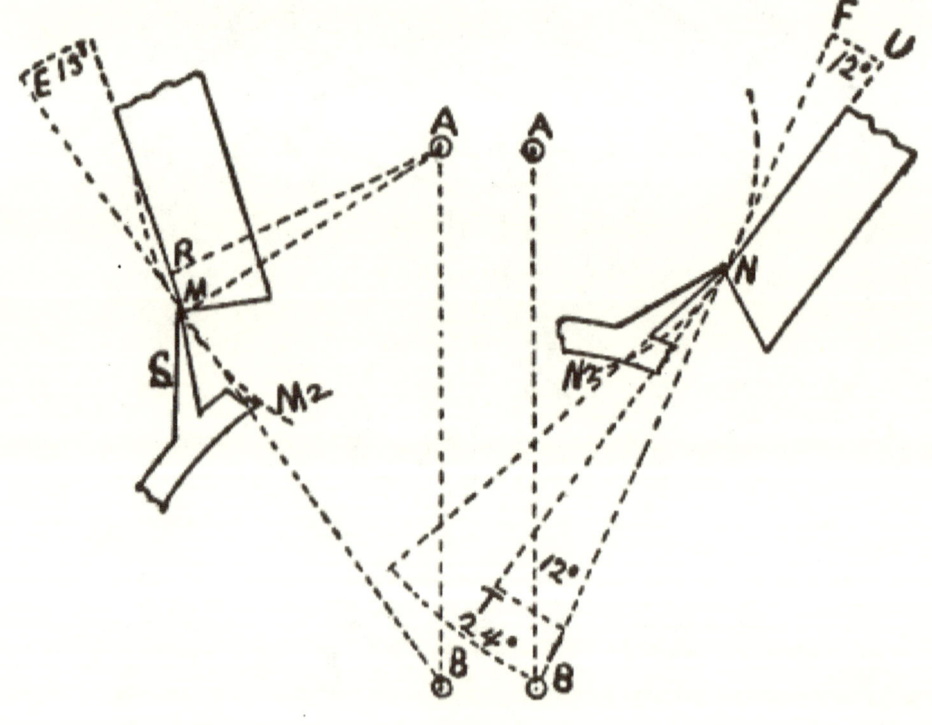

FIG. 5.

Pallets with draw are more difficult to unlock than those without it, this is in the nature of a fault, but whenever there are two faults we must choose the less. The rate of the watch will suffer less on account of the recoil introduced than it would were the locking faces arcs of circles struck from the pallet center, in which case the guard point would often remain against the roller. The draw should be as light as possible consistent with safety of action; some writers allow 15° on the engaging and 12° on the disengaging pallet; others again allow 12° on each, which we deem sufficient. The draw is measured from the locking edges M and N, Fig. 5. The locking planes *when locked* are inclined 12° from EB, and FB. In the case of the engaging pallet it inclines toward the center A. The draw is produced on account of MA being longer than RA, consequently, when power is applied to the scape tooth S, the pallet is drawn into the wheel. The disengaging pallet inclines in the same direction but away from the center A; the reason is obvious from the former explanation. Some people imagine that the greater the incline on the locking edge of the escape teeth, the stronger the draw would be. This is not the case, but it is certainly necessary that the point of the tooth alone should touch the pallet. From this it follows that the angle on the teeth must be greater than on the pallets; examine the disengaging pallet in Fig. 5, as it is from this pallet that the inclination of the teeth must be determined, as in the case of the engaging pallet the motion is toward the line of centers AB, and therefore *away* from the tooth, which partially explains why some people advocate 15° draw for this pallet. As illustrated in the case of the disengaging pallet, however, the motion is also towards the line of centers AB, and *towards* the tooth as well, all of which will be seen by the dotted circles MM2 and NN2, representing the paths of the pallets. It will be noticed that UNF and BNB are opposite and equal angles of 12°. For practical reasons, from a manufacturing standpoint, the angle on the tooth is made just twice the amount, namely 24°; we could make it a little less or a little more. If we made it less than 20° too great a surface would be in contact with the jewel, involving greater friction in unlocking and an inefficient draw, but in the case of an English lever with such an arrangement we could do with less drop, which advantage would be too dearly bought; or if the angle is made over 28°, the point or locking edge of the tooth would rapidly become worn in case of a brass wheel. Also in an English lever more drop would be required.

THE LOCK.

The Lock. — What we have said in regard to drop also applies to the lock, which should be as small as possible, consistent with perfect safety. The greater the drop the deeper must be the lock; 1½° is the angle generally allowed for the lock, but it is obvious that in a large escapement it can be less.

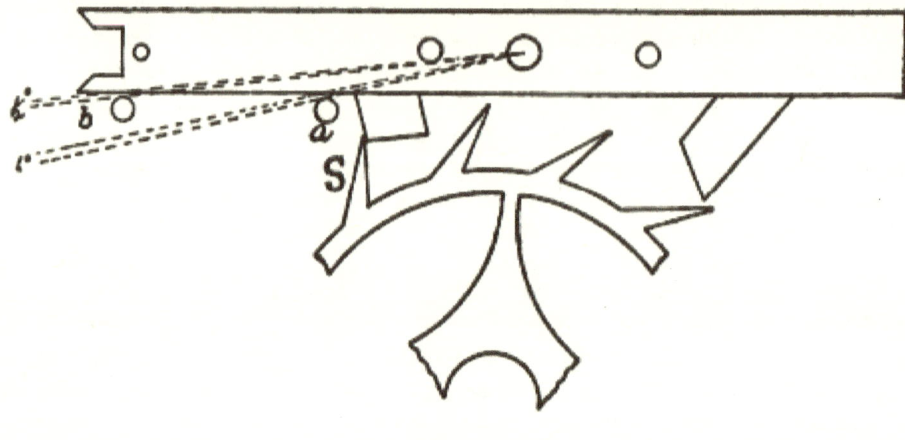

FIG. 6.

THE RUN.

The Run.—The run or, as it is sometimes called, "the slide," should also be as light as possible; from ¼° to ½° is sufficient. It follows then, the bankings should be as close together as possible, consistent with requisite freedom for escaping. Anything more than this increases the angular connection of the balance with the escapement, which directly violates the theory under which it is constructed; also, a greater amount of work will be imposed upon the balance to meet the increased unlocking resistance, resulting in a poor motion and accurate time will be out of the question. It will be seen that those workmen who make a practice of opening the banks, "to give the escapement more freedom" simply jump from the frying pan into the fire. The bankings should be as far removed from the pallet center as possible, as the further away they are pitched the less run we require, according to angular measurement. Figure 6 illustrates this fact; the tooth S has just dropped on the engaging pallet, but the fork has not yet reached the bankings. At *a* we have 1° of run, while if placed at *b* we would only have ½° of run, but still the same freedom for escaping, and less unlocking resistance.

The bankings should be placed towards the acting end of the fork as illustrated, as in case the watch "rebanks" there would be more strain on the lever pivots if they were placed at the other end of the fork.

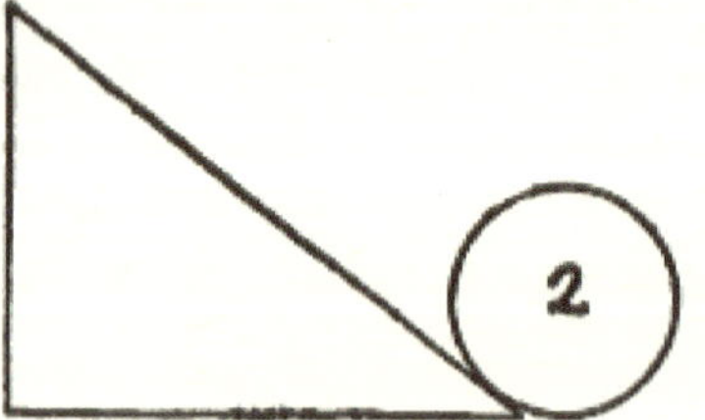

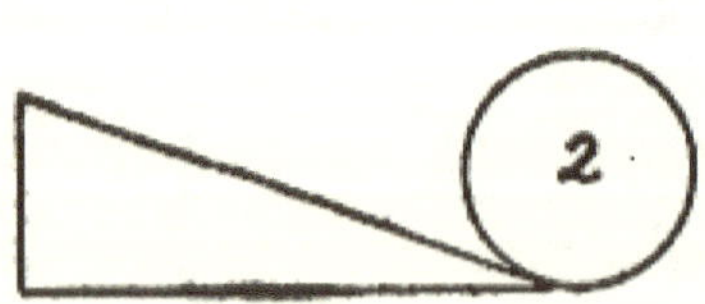

FIG. 7.

THE LIFT.

The Lift. — The lift is composed of the actual lift on the teeth and pallets and the lock and run. We will suppose that from drop to drop we allow 10°; if the lock is 1½° then the actual lift by means of the inclined planes on teeth and pallets will be 8½°. We have seen that a small lifting angle is advisable, so that the vibrations of the balance will be as free as possible. There are other reasons as well. Fig. 7 shows two inclined planes; we desire to lift the weight 2 a distance equal to the angle at which the planes are inclined; it will be seen at a glance that we will have less friction by employing the smaller incline, whereas with the larger one the motive power is employed through a greater distance on the object to be moved. The smaller the angle the more energetic will the movement be; the grinding of the angles and fit of the pivots, etc., also increases in importance. An actual lift of 8½° satisfies the conditions imposed very well. We have before seen that both on account of the unlocking and the lifting leverage of the pallet arms, it would be advisable to make them narrow both in the equidistant and circular escapement. We will now study the question from the standpoint of the lift, in so far as the wheel is concerned.

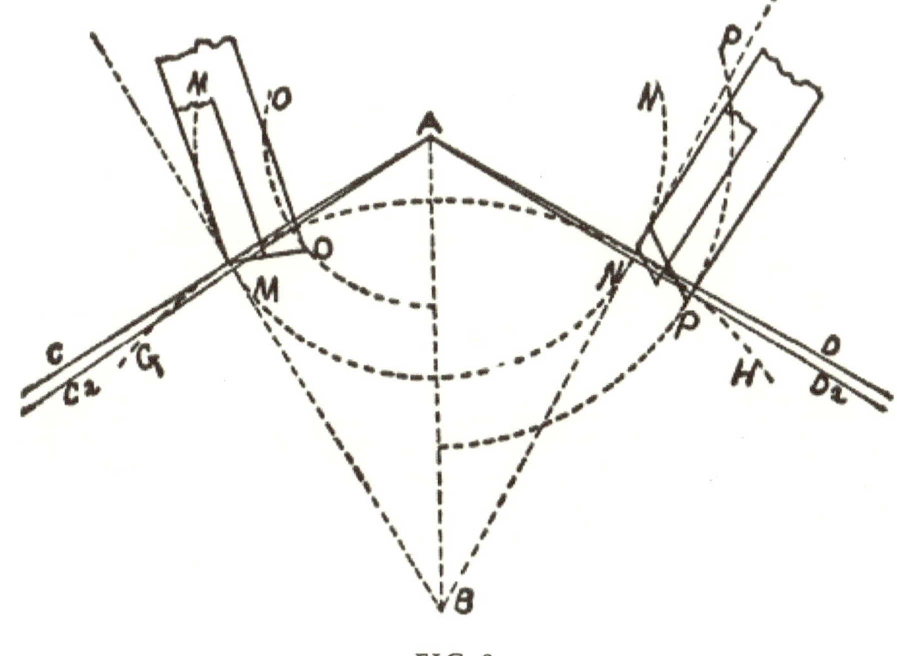

FIG. 8.

It is self-evident that a narrow pallet requires a wide tooth, and a wide pallet a narrow or thin tooth wheel; in the ratchet wheel we have a metal point passing over a jeweled plane. The friction is at its minimum, because there is less adhesion than with the club tooth, but we must emphasize the fact that we require a greater angle in proportion on the pallets in this escapement than with the narrow pallets and wider tooth. This seems to be a point which many do not thoroughly comprehend, and we would advise a close study of Fig. 8, which will make it perfectly clear, as we show both a wide and a narrow pallet. GH, represents the primitive, which in this figure is also the real diameter of the escape wheel. In measuring the lifting angles for the pallets, our starting point is *always* from the tangents AC and AD. The tangents are straight lines, but the wheel describes the circle GH, therefore they must deviate from one another, and the closer to the center A the discharging edge of the engaging pallet reaches, the greater does this difference become; and in the same manner the further the discharging edge of the disengaging pallet is from the center A the greater it is. This shows that the loss is greater in the equidistant than in the circular escapement. After this we will designate this difference as the "loss." In order to illustrate it more plainly we show the widest pallet—the English—in equidistant form. This gives another reason why the English lever should only be made with circular pallets, as we have seen that the wider the pallet the greater the loss. The loss is measured at the intersection of the path of the discharging edge OO, with the circle G H, and is shown through AC2, which intersects these circles at that point. In the case of the disengaging pallet, PP illustrates the path of the discharging edge; the loss is measured as in the preceding case where GH is intersected as shown by AD2. It amounts to a different value on each pallet. Notice the loss between C and C2, on the engaging, and D and D2 on the disengaging pallet; it is greater on the engaging pallet, so much so that it amounts to 2°, which is equal to the entire lock; therefore if 8½° of work is to be accomplished through this pallet, the lifting plane requires an angle of 10½° struck from AC.

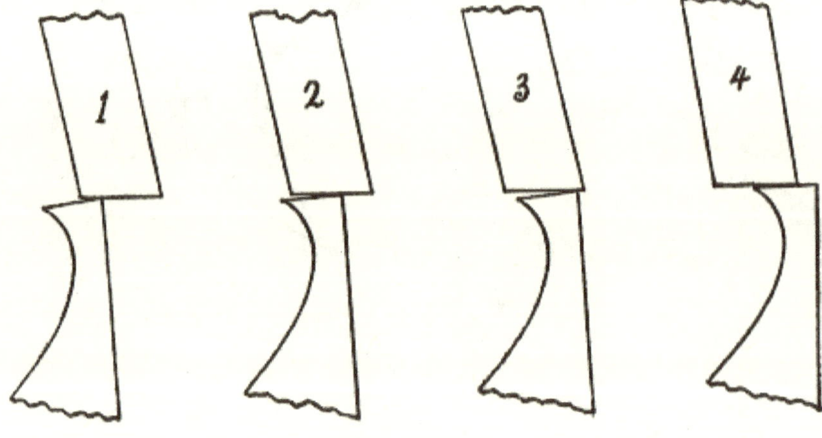

FIG. 9.

Let us now consider the lifting action of the club tooth wheel. This is decidedly a complicated action, and requires some study to comprehend. In action with the engaging pallet the wheel moves *up*, or in the direction of the motion of the pallets, but on the disengaging pallet it moves *down*, and in a direction opposite to the pallets, and the heel of the tooth moves with greater velocity than the locking edge; also in the case of the engaging pallet, the locking edge moves with greater velocity than the discharging edge; in the disengaging pallet the opposite is the case, as the discharging edge moves with greater velocity than the locking. These points involve factors which must be considered, and the drafting of a correct action is of paramount importance; we therefore show the lift as it is accomplished in four different stages in a good action. Fig. 9 illustrates the engaging, and Fig. 10 the disengaging pallet; by comparing the figures it will be noticed that the lift takes place on the point of the tooth similar to the English, until the discharging edge of the pallet has been passed, when the heel gradually comes into play on the engaging, but more quickly on the disengaging pallet.

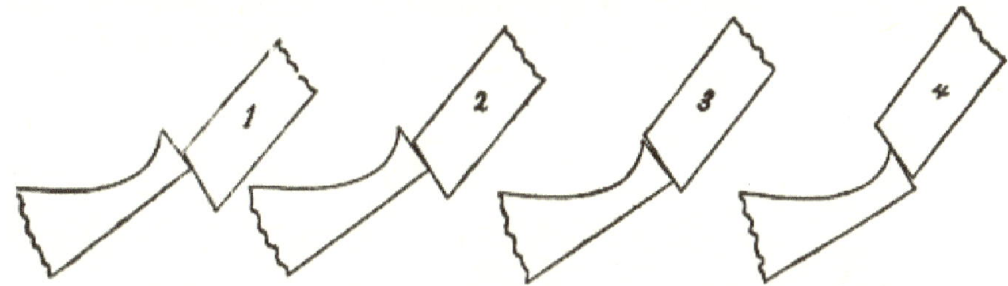

FIG. 10.

We will also notice that during the first part of the lift the tooth moves faster along the engaging lifting plane than on the disengaging; on pallets 2 and 3 this difference is quite large; towards the latter part of the lift the action becomes quicker on the disengaging pallet and slower on the engaging.

To obviate this difficulty some fine watches, notably those of A. Lange & Sons, have convex lifting planes on the engaging and concave on the disengaging pallets; the lifting planes on the teeth are also curved. See Fig. 11. This is decidedly an ingenious arrangement, and is in strict accordance with scientific investigation. We should see many fine watches made with such escapements if the means for producing them could fully satisfy the requirements of the scientific principles involved.

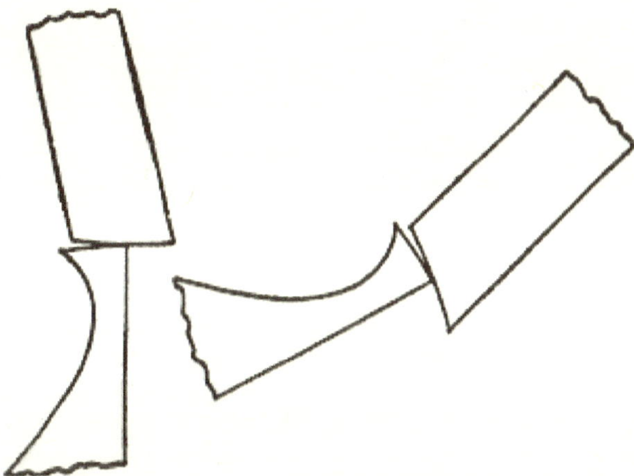

FIG. 11.

The distribution of the lift on tooth and pallet is a very important matter; the lifting angle on the tooth must be *less* in proportion to its width than it is on the pallet. For the sake of making it perfectly plain, we illustrate what should not be made; if we have 10½° for width of tooth and pallet, and take half of it for a tooth, and the other half for the pallet, making each of them 5¼° in width, and suppose we have a lifting of 8½° to distribute between them, by allowing 4¼° on each, the lift would take place as shown in Fig. 12, which is a very unfavorable action. The edge of the engaging pallet scrapes on the lifting plane of the tooth, yet it is astonishing to find some otherwise very fine watches being manufactured right along which contain this fault; such watches can be stopped with the ruby pin in the fork and the engaging pallet in action, nor would they start when run down as soon as the crown is touched, no matter how well they were finished and fitted.

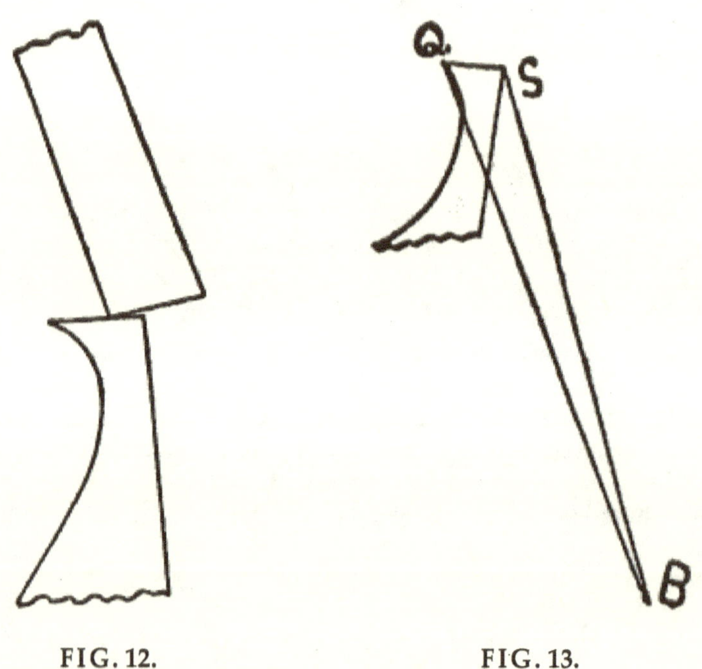

FIG. 12.   FIG. 13.

The lever lengths of the club tooth are variable, while with the ratchet they are constant, which is in its favor; in the latter it would always be as SB, Fig. 13. This is a shorter lever than QB, consequently more powerful, although the greater velocity is at Q, which only comes into action after the inertia of wheel and pallets has been overcome, and when the greatest momentum during contact is reached. SB is the primitive radius of the club tooth wheel, but both primitive and *real* radius of the ratchet wheel. The distance of centers of wheel and pallet will be alike in both cases; also the lockings will be the same distance apart on both pallets; therefore, when horologists, even if they have worldwide reputations, claim that the club tooth has an advantage over the ratchet because it begins the lift with a shorter lever than the latter, it does not make it so. We are treating the subject from a purely horological standpoint, and neither patriotism or prejudice has anything to do with it. We wish to sift the matter thoroughly and arrive at a just conception of the merits and defects of each form of escapement, and show *reasons* for our conclusions.

Anyone who has closely followed our deductions must see that in so far as the wheel is concerned the ratchet or English wheel has several points in its favor. Such a wheel is inseparable from a wide pallet; but we have seen that a narrower pallet is advisable; also as little drop and lock as possible; clearly, we must effect a compromise. In other words, so far the balance of our reasoning is in favor of the club tooth escapement and to effect an intelligent division of angles for tooth, pallet and lift is one of the great questions which confronts the intelligent horologist.

Anyone who has ever taken the pains to draw pallet and tooth with different angles, through every stage of the lift, with both wide and narrow pallets and teeth, in circular and equidistant escapements, will have received an eye-opener. We strongly advise all our readers who are practical workmen to try it after studying what we have said. We are certain it will repay them.

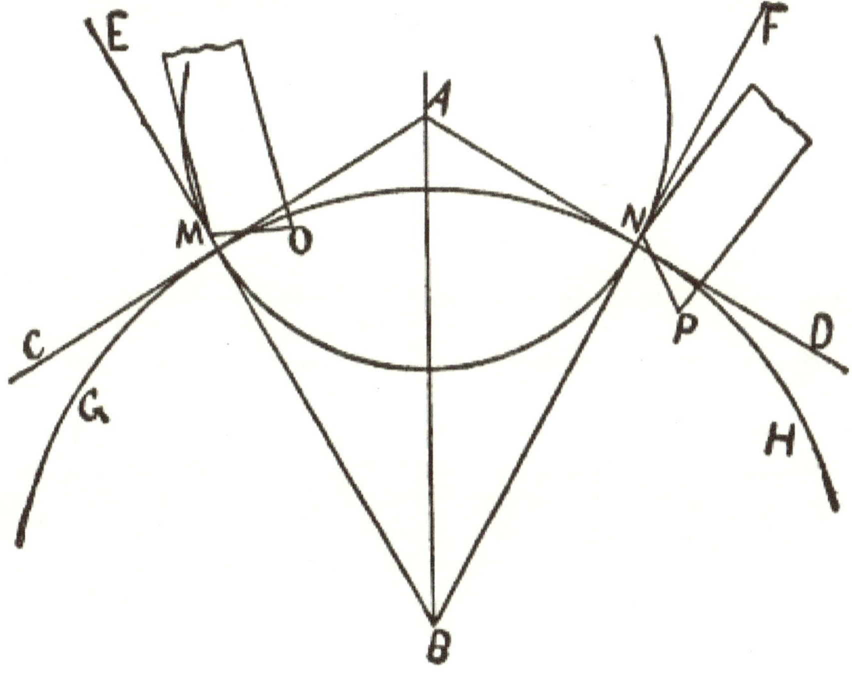

FIG. 2.

THE CENTER DISTANCE OF WHEEL AND PALLETS.

The Center Distance of Wheel and Pallets. The direction of pressure of the wheel teeth should be through the pallet center by drawing the tangents AC and AD, Fig. 2 to the primitive circle GH, at the intersection of the angle FBE. This condition is realized in the equidistant pallet. In the circular pallet, Fig. 3, this condition cannot exist, as in order *to lock* on a tangent the center distance should be *greater* for the engaging and *less* for the disengaging pallet, therefore watchmakers aim to go between the two and plant them as before specified at A.

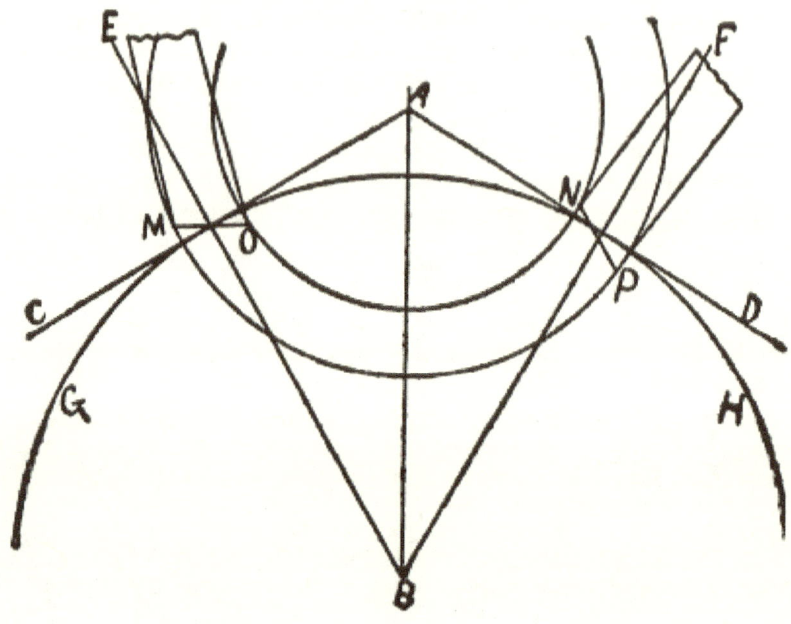

FIG. 3.

When planted on the tangents the unlocking resistance will be less and the impulse transmitted under favorable conditions, especially so in the circular, as the direction of pressure coincides (close to the center of the lift), with the law of the parallelogram of forces.

It is *impossible* to plant pallets on the tangents in very small escapements, as there would not be enough room for a pallet arbor of proper strength, nor will they be found planted on the tangents in the medium size escapement with a long pallet arbor, nor in such a one with a very wide tooth (see Fig. 4) as the heel would come so close to the center A, that the solidity of pallets and arbor would suffer. We will give an actual example. For a medium sized escape wheel with a primitive diameter of 7.5 mm., the center distance AB is 4.33 mm. By using 3° of a lifting angle on the teeth, the distance from the heel of the tooth to the pallet center will be .4691 mm.; by allowing .1 mm. between wheel and pallet and .15 mm. for stock on the pallets we find we will have a pallet arbor as follows: .4691 − (.1 + .15) × 2 = .4382 mm. It would not be practicable to make anything smaller.

It behooves us now to see that while a narrow pallet is advisable a very wide tooth is not; yet these two are inseparable. Here is another case for a compromise, as, unquestionably the pallets ought to be planted on the tangents. There is no difficulty about it in the English lever, and we have shown in our example that a judiciously planned club tooth escapement of medium size can be made with the center distance properly planted.

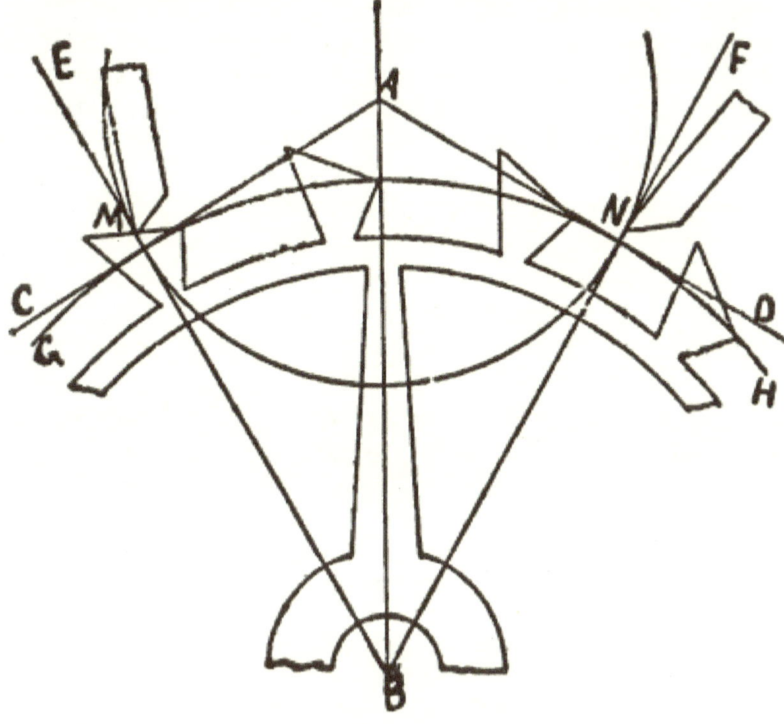

FIG. 4.

When considering the center distance we must of necessity consider the widths of teeth and pallets and their lifting angles. We are now at a point in which no watchmaker of intelligence would indicate one certain division for these parts and claim it to be "the best." It is always those who do not thoroughly understand a subject who are the first to make such claims. We will, however, give our opinion within certain limits. The angle to be divided for tooth and pallet is 10½°. Let us divide it by 2, which would be the most natural thing to do, and examine the problem. We will have 5¼° each for width of tooth and pallet. We *must* have a smaller lifting angle on the tooth than on the pallet, but the wider the tooth the greater should its lifting angle be. It would not be mechanical to make the tooth wide and the lifting angle small, as the lifting plane on the pallets would be too steep on account of being narrow. A lifting angle on the tooth which would be *exactly* suitable for a given circular, would be *too great* for a given equidistant pallet. It follows, therefore, taking 5¼° as a width for the tooth, that while we could employ it in a fair sized escapement with equidistant pallets, we could not do so with circular pallets and still have the latter pitched on the tangents. We see the majority of escapements made with narrower teeth than pallets, and for a very good reason.

In the example previously given, the 3° lift on the tooth is well adapted for a width of 4½°, which would require a pallet 6° in width. The tooth, therefore, would be ¾ the width of pallets, which is very good indeed.

From what we have said it follows that a large number of pallets are not planted on the tangents at all. We have never noticed this question in print before. Writers generally seem to, in fact do, assume that no matter how large or small the escapement may be, or how the pallets and teeth are divided for width and lifting angle, no difficulty will be found in locating the pallets on the tangents. Theoretically there is no difficulty, but in practice we find there is.

EQUIDISTANT VS. CIRCULAR.

Equidistant vs. Circular. At this stage we are able to weigh the circular against the equidistant pallet. In beginning this essay we had to explain the difference between them, so the reader could follow our discussion, and not until now, are we able to sum up our conclusions.

The reader will have noticed that for such an important action as the lift, which supplies power to the balance, the circular pallet is favored from every point of view. This is a very strong point in its favor. On the other hand, the unlocking resistance being less, and as nearly alike as possible on both pallets in the equidistant, it is a question if the total vibration of the balance will be greater with the one than the other, although it will receive the impulse under better conditions from the circular pallet; but it expends more force in unlocking it. Escapement friction plays an important role in the position and isochronal adjustments; the greater the friction encountered the slower the vibration of the balance. The friction should be constant. In unlocking, the equidistant comes nearer to fulfilling this condition, while during the lift it is more nearly so in the circular. The friction in unlocking, from a timing standpoint, overshadows that of the impulse, and the tooth can be a little wider in the equidistant than the circular escapement with the pallet properly planted. Therefore for the *finest* watches the equidistant escapement is well adapted, but for anything less than that the circular should be our choice.

THE FORK AND ROLLER ACTION.

The Fork and Roller Action. While the lifting action of the lever escapement corresponds to that of the cylinder, the fork and roller action corresponds to the impulse action in the chronometer and duplex escapements.

Our experience leads us to believe that the action now under consideration is but imperfectly understood by many workmen. It is a complicated action, and when out of order is the cause of many annoying stoppages, often characterized by the watch starting when taken from the pocket.

The action is very important and is generally divided into impulse and safety action, although we think we ought to divide it into three, namely, by adding that of the unlocking action. We will first of all consider the impulse and unlocking actions, because we cannot intelligently consider the one without the other, as the ruby pin and the slot in the fork are utilized in each. The ruby pin, or strictly speaking, the "impulse radius," is a lever arm, whose length is measured from the center of the balance staff to the face of the ruby pin, and is used, firstly, as a power or transmitting lever on the acting or geometrical length of the fork (*i. e.*, from the pallet center to the beginning of the horn), and which at the moment is a resistance lever, to be utilized in unlocking the pallets. After the pallets are unlocked the conditions are reversed, and we now find the lever fork, through the pallets, transmitting power to the balance by means of the impulse radius. In the first part of the action we have a short lever engaging a longer one, which is an advantage. See Fig. 14, where we have purposely somewhat exaggerated the conditions. A'X represents the impulse radius at present under discussion, and AW the acting length of the fork. It will be seen that the shorter the impulse radius, or in other words, the closer the ruby pin is to the balance staff and the longer the fork, the easier will the unlocking of the pallets be performed, but this entails a great impulse angle, for the law applicable to the case is, that the angles are in the inverse ratio to the radii. In other words, the shorter the radius, the greater is the angle, and the smaller the angle the greater is the radius. We know, though, that we must have as small an impulse angle as possible in order that the balance should be highly detached. Here is one point in favor of a short impulse radius, and one against it. Now, let us turn to the impulse action. Here we have the long lever AW acting on a short one, A'X, which is a disadvantage. Here, then, we ought to try and have a short lever acting on a long one, which would point to a short fork and a great impulse radius. Suppose AP, Fig. 14, is the length of fork, and A'P is the impulse radius; here, then, we favor the impulse, and it is directly in accordance with the theory of the free vibration of the balance, for, as before stated, the longer the radius the smaller the angle. The action at P is also closer to the line of centers than it is at W, which is another advantage.

We will notice that by employing a large impulse angle, and consequently a short radius, the intersection *m* of the two circles *ii* and *cc* is very *safe*, whereas, with the conditions reversed in favor of the impulse action, the intersection at *k* is more delicate. We have now seen enough to appreciate the fact that we favor one action at the expense of another.

By having a lifting angle on pallet and tooth of 8½°, a locking angle of 1½°, and a run of ½°, we will have an angular movement of the fork of 8½ + 1½ + ½ = 10½°.

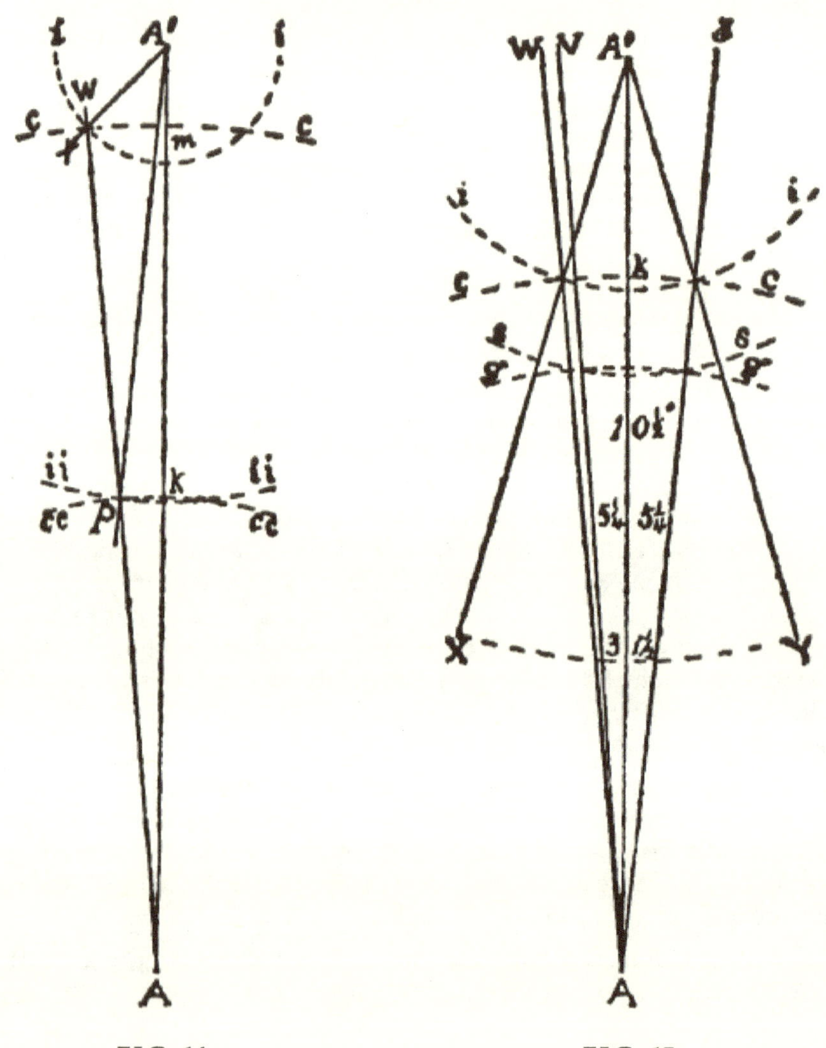

FIG. 14.   FIG. 15.

Writers generally only consider the movement of the fork from drop to drop on the pallets, but we will be thoroughly practical in the matter. With a total motion of the fork of 10½° (JAW, Fig. 15), one-half, or 5¼° will be performed on each side of the line of centers. We are at liberty to choose any impulse angle which we may prefer; 3 to 1 is a good proportion for an ordinary well-made watch. By employing it, the angle XA′Y would be equal to 31½°. The radius A′X Fig. 16, is also of the same proportion, but the angle AA′X is greater because the fork angle WAA′ is greater than the same angle in Fig. 15. We will notice that the intersection k is much smaller in Fig. 15 than in Fig. 16. The action in the latter begins much further from the line of centers than in the former and outlines an action which should not be made.

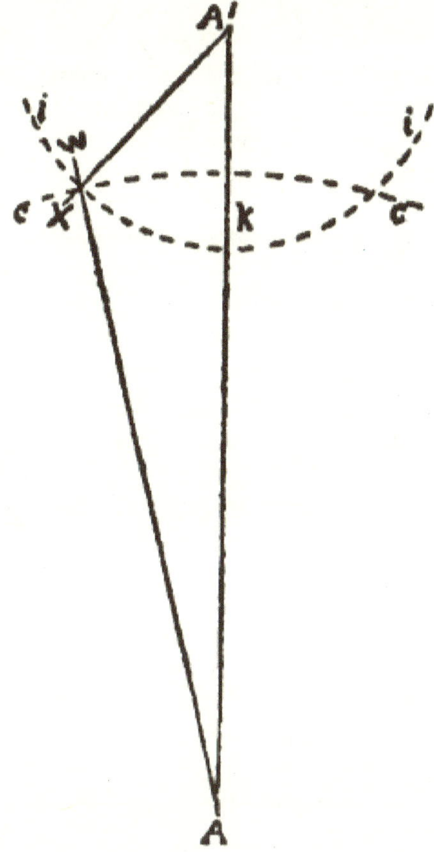

FIG. 16.

To come back to the impulse angle, some might use a proportion of 3.5, 4 or even 5 to 1, while others for the finest of watches would only use 2.75 to 1. By having a total vibration of the balance of 1½ turns, which is equal to 540° a fork angle of 10° and a proportion of 2.75 for the impulse angle which would be equal to 10 × 2.75 = 27.5°. The *free* vibration of the balance, or as this is called, "the supplemental arc," is equal to 540° − 27.5° = 512.50°, while with a proportion of 5 to 1, making an impulse angle of 50°, it would be equal to 490°. To sum up, the finer the watch the lower the proportion, the closer the action to the line of centers, the smaller the friction. On account of leverage the more difficult the unlocking but the more energetic the impulse when it does occur. The velocity of the ruby pin at P; Fig. 14, is much greater than at W, consequently it will not be overtaken as soon by the fork as at W. The velocity of the fork at the latter point is greater than at P; the intersection of *ii* and *cc* is also not as great; therefore the lower the proportion the finer and more exact must the workmanship be.

We will notice that the unlocking action has been overruled by the impulse. The only point so far in which the former has been favored is in the diminished action before the line of centers, as previously pointed out at P, Fig. 14.

We will now consider the width of the ruby pin and to get a good insight into the question, we will study Fig. 17. A is the pallet center, A' the balance center, the line AA' being the line of centers; the angle WAA equals half the total motion of the fork, the other half, of course, taking place on the opposite side of the center line. WA is the *center* of the fork when it rests against the bank. The angle AA'X represents half the impulse angle; the other half, the same as with the fork, is struck on the other side of the center line. At the point of intersection of these angles we will draw cc from the pallet center A, which equals the acting length of the fork, and from the balance center we will draw ii, which equals the *theoretical* impulse radius; some writers use it as the *real* radius. The wider the ruby pin the greater will the latter be, which we will explain presently.

The ruby pin in entering the fork must have a certain amount of freedom for action, from 1 to 1¼°. Should the watch receive a jar at the moment the guard point enters the crescent or passing hollow in the roller, the fork would fly against the ruby pin. It is important that the angular freedom between the fork and ruby pin at the moment it enters into the slot be *less* than the total locking angle on the pallets. If we employ a locking angle of 1½° and ½° run, we would have a total lock on the pallets of 2°. By allowing 1¼° of freedom for the ruby pin at the moment the guard point enters the crescent, in case the fork should strike the face of the ruby pin, the pallets will still be locked ¾° and the fork drawn back against the bankings through the draft angle.

We will see what this shake amounts to for a given acting length of fork, which describes an arc of a circle, therefore the acting length is only the radius of that circle and must be multiplied by two in order to get the diameter. The acting length of fork = 4.5 mm., what is the amount of shake when the ruby pin passes the acting corner? 4.5 × 2 × 3.1416 ÷ 360° = .0785 × 1.25 = .0992 mm. The shake of the ruby pin in the slot of the fork must be as slight as possible, consistent with perfect freedom of action. It varies from ¼° to ½°, according to length of fork and shape of ruby pin. A square ruby pin requires more shake than any other kind; it enters the fork and receives the impulse in a diagonal direction on the jewel, in which position it is illustrated at Z, Fig. 20. This ruby pin acts on a knife edge, but for all that the engaging friction during the unlocking action is considerable.

Our reasoning tells us it matters not if a ruby pin be wide or narrow, it must have *the same* freedom in passing the acting edge of the fork, therefore, to have the impulse radius on the point of intersection of A'X with AW, Fig. 17, we would require a *very* narrow ruby pin. With 1° of freedom at the edge, and ½° in the slot, we could only have a ruby pin of a width of 1½°. Applying it to the preceding example it would only have an actual width of .0785 × 1.5 = .1178 mm., or the size of an ordinary balance pivot. At *n*, Fig. 17, we illustrate such a ruby pin; the theoretical and real impulse radius coincide with one another. The intersection of the circle ii and cc is very slight, while the friction in unlocking begins within 1° of half the total movement of the fork from the line of centers; to illustrate, if the angular motion is 11° the ruby pin under discussion will begin action 4½° before the line of centers, being an engaging, or "uphill" friction of considerable magnitude.

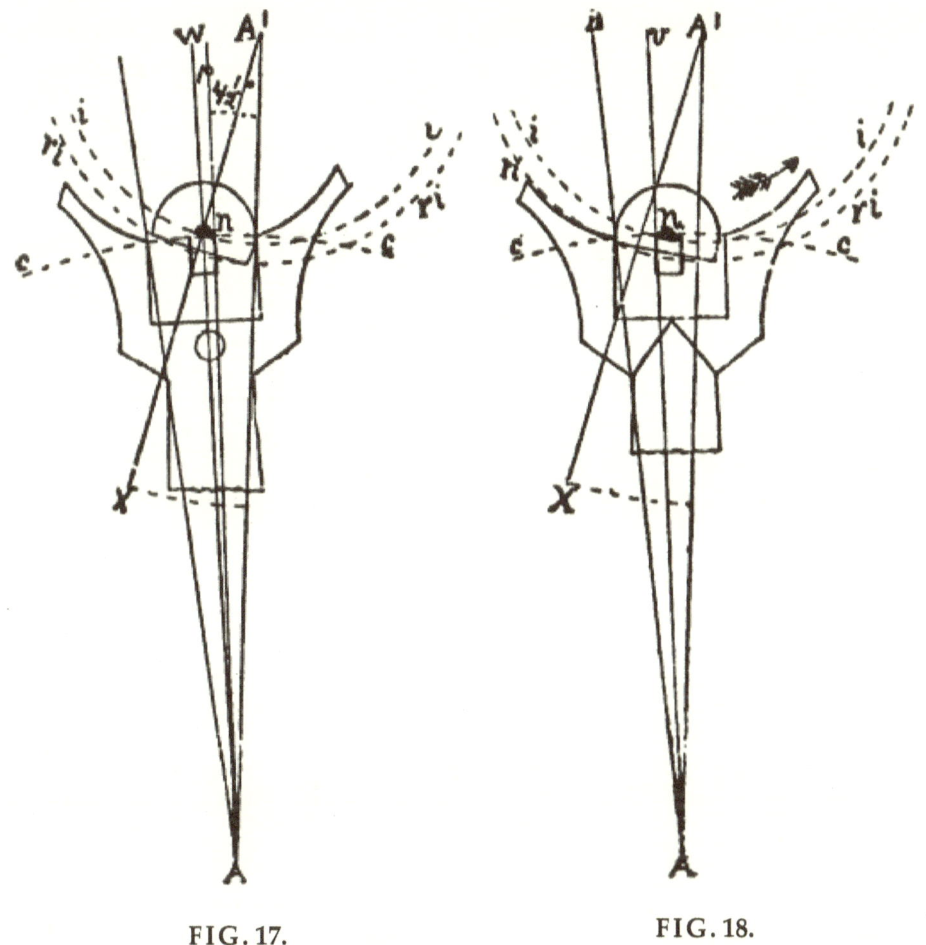

FIG. 17.   FIG. 18.

The intersection with the fork is also much less than with the wider ruby pin, making the impulse action very delicate. On the other hand the widest ruby pin for which there is any occasion is one beginning the unlocking action on the line of centers, Fig. 17; this entails a width of slot equal to the angular motion of the fork. We see here the advantage of a wide ruby pin over a narrow one in the unlocking action. Let us now examine the question from the standpoint of the impulse action.

Fig. 18 illustrates the moment the impulse is transmitted; the fork has been moved in the direction of the arrow by the ruby pin; the escapement has been unlocked and the opposite side of the slot has just struck the ruby pin. The exact position in which the impulse is transmitted varies with the locking angle, the width of ruby pin, its shake in the slot, the length of fork, its weight, and the velocity of the ruby pin, which is determined by the vibrations of the balance and the impulse radius.

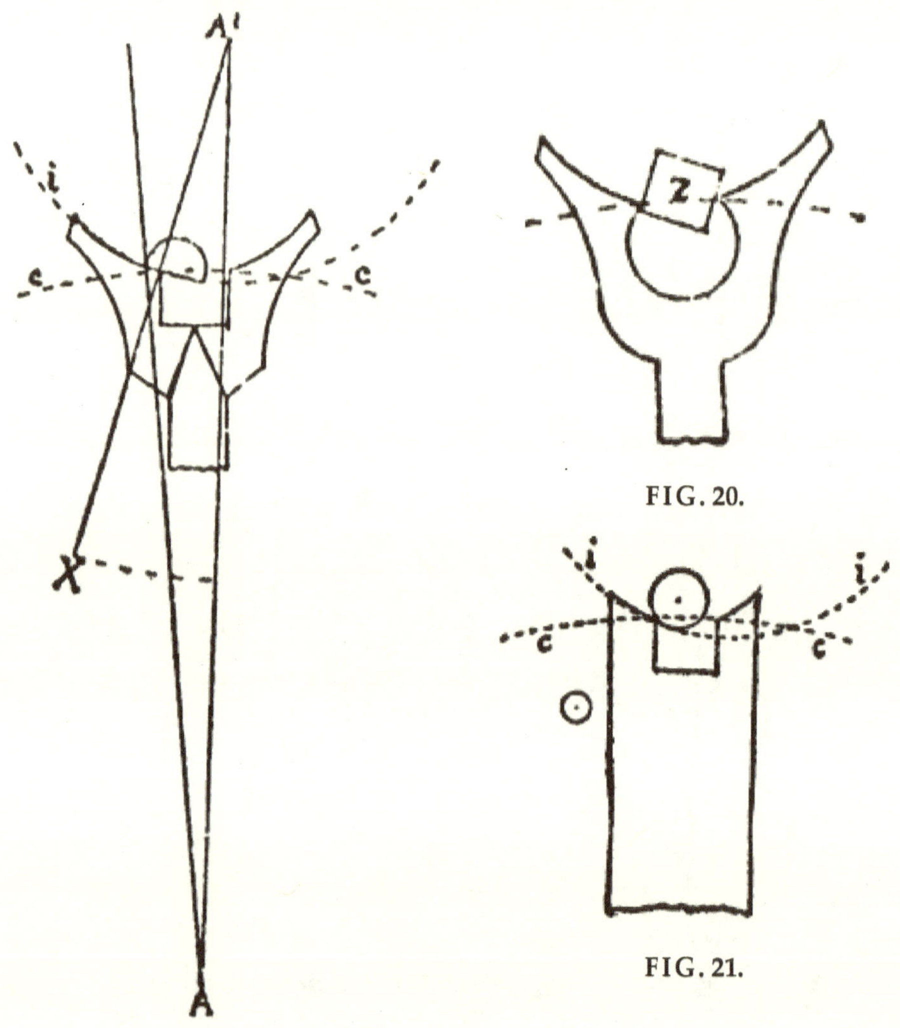

FIG. 20.

FIG. 21.

FIG. 19.

In an escapement with a total lock of 1¾° and 1¼ of shake in the slot, theoretically, the impulse would be transmitted 2° from the bankings. The narrow ruby pin n receives the impulse on the line v, which is closer to the line of centers than the line u, on which the large ruby pin receives the impulse. Here then we have an advantage of the narrow ruby pin over a wide one; with a wider ruby pin the balance is also more liable to rebank when it takes a long vibration. Also on account of the greater angle at which the ruby pin stands to the slot when the impulse takes place, the *drop* of the fork against the jewel will amount to more than its shake in the slot (which is measured when standing on the line of centers). On this account some watches have slots dovetailed in form, being wider at the bottom, others have ruby pins of this form. They require very exact execution; we think we can do without them by judiciously selecting a width of ruby pin between the two extremes. We would choose a ruby pin of a width equal to half the angular motion of the fork. There is an ingenious arrangement of fork and roller which aims to, and partially does, overcome the difficulty of choosing between a wide and narrow ruby pin, it is known as the Savage pin roller escapement. We intend to describe it later.

If the face of the ruby pin were planted on the theoretical impulse radius *ii*, Fig. 19, the impulse would end in a butting action as shown; hence the great importance of distinguishing between the theoretical and real impulse radius and establishing a reliable data from which to work. We feel that these actions have never been properly and thoroughly treated in simple language; we have tried to make them plain so that anyone can comprehend them with a little study.

Three good forms of ruby pins are the triangular, the oval and the flat faced; for ordinary work the latter is as good as any, but for fine work the triangular pin with the corners slightly rounded off is preferable.

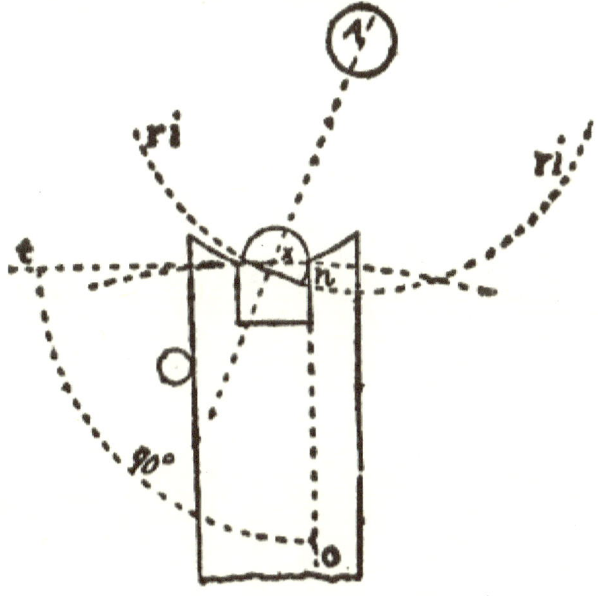

FIG. 22.

English watches are met with having a cylindrical or round ruby pin. Such a pin should never be put into a watch. The law of the parallelogram of forces is completely ignored by using such a pin; the friction during the unlocking and impulse actions is too severe, as it is, without the addition of so unmechanical an arrangement. Fig. 21 illustrates the action of a round ruby pin; *ii* is the path of the ruby pin; *cc* that of the acting length of the fork. It is shown at the moment the impulse is transmitted. It will be seen that the impact takes place *below* the center of the ruby pin, whereas it should take place at the center, as the motion of the fork is *upwards* and that of the ruby pin *downwards* until the line of the centers has been reached. The same rule applies to the flat-faced pin and it is important that the right quantity be ground off. We find that $\frac{3}{7}$ is approximately the amount which should be ground away. Fig. 22 illustrates the fork standing against the bank. The ruby pin touches the side of the slot but has not as yet begun to act; *ri* is the real impulse circle for which we allow 1¼° of freedom at the acting edge of the fork; the face of the ruby pin is therefore on this line. The next thing to do is to find the center of the pin. From the side *n* of the slot we construct the right angle *o n t*; from *n*, we transmit ½ the width of the pin, and plant the center *x* on the line *n t*. We can have the center of the pin slightly below this line, but in no case above it; but if we put it below, the pin will be thinner and therefore more easily broken.

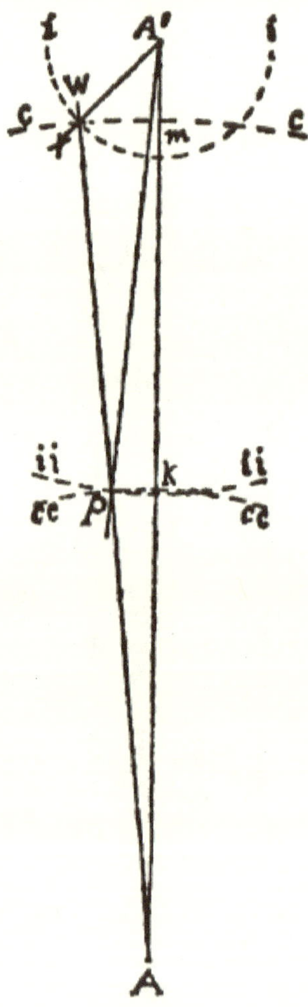

FIG. 14.

THE SAFETY ACTION.

The Safety Action. Although this action is separate from the impulse and unlocking actions, it is still very closely connected with them, much more so in the single than in the double roller escapement. If we were to place the ruby pin at X, Fig. 14, we could have a much smaller roller than by placing it at P. With the small roller the safety action is more secure, as the intersection at m is greater than at k. It is not as liable to "butt" and the friction is less when the guard point is thrown against the small roller. Suppose we take two rollers, one with a diameter of 2.5 mm., the other just twice this amount, of 5 mm. By having the guard radius and pressure the same in each case, if the guard point touched the larger roller it would not only have twice, but four times more effect than on the smaller one. We will notice that the smaller the impulse angle the larger the roller, because the ruby pin is necessarily placed farther from the center. The position of the ruby pin should, therefore, govern the size of the roller, which should be as small as possible. There should only be enough metal left between the circumference of the roller and the face of the jewel to allow for a crescent or passing hollow of sufficient depth and an efficient setting for the jewel. For this reason, as well as securing the correct impulse radius and therefore angle, when replacing the ruby pin, and having it set securely and mechanically in the roller, it is necessary that the pin and the hole in the roller be of the same form, and a good fit. Fig. 23 illustrates the difference in size of rollers. In the smaller one the conditions imposed are satisfied, while in the larger one they are not. In the single roller the safety action is at the mercy of the impulse and pallet angles. We have noticed that in order to favor the impulse we require a large roller, and for the safety action a small one, therefore escapements made on fine principles are supplied with two rollers, one for each action.

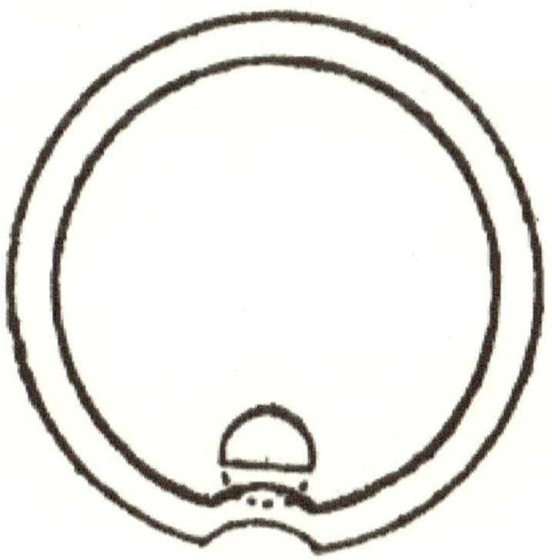

FIG. 23.

It may be well to say that in our opinion a proportion between the fork and impulse angles in 10° pallets of 3 or 3½ to 1, *depending* upon the size of the escapement, is the lowest which should be made in single roller. We have seen them in proportions of 2 to 1 in single roller — a scientific principle foolishly applied — resulting in an action entirely unsatisfactory.

When the guard point is pressed against the roller the escape tooth must still rest on the locking face of the pallet; if the total lock is 2°, by allowing 1¼° freedom for the guard point between the bank and the roller the escapement will still be locked ¾°. How much this shake actually amounts to depends upon the guard radius. Suppose this to be 4 mm., then the freedom would equal 4 × 2 × 3.1416 ÷ 360 × 1.25 = .0873 mm.

THE CRESCENT.

The Crescent in the roller must be large and deep enough so it will be impossible for the guard point to touch in or on the corners of it; at the same time it must not be too large, as it would necessitate a longer horn on the fork than is necessary.

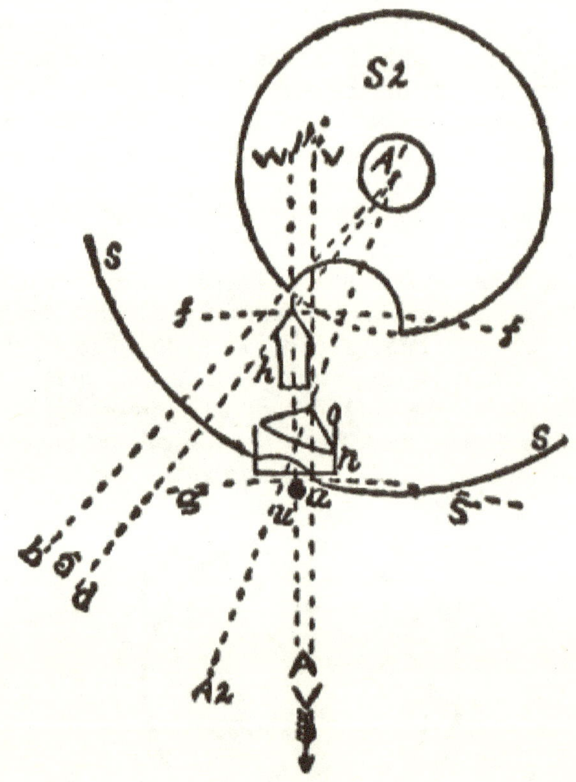

FIG. 24.

Fig. 24 shows the slot *n* of the fork standing at the bank. The ruby pin *o* touches it, but has not as yet acted on it; *s s* illustrates a single roller, while S2 illustrates the safety roller for a double roller escapement. In order to find the dimensions of the crescent in the single roller we must proceed as follows: WA is in the center of the fork when it rests against the bank, and is, therefore, one of the sides of the fork angle, and is drawn from the pallet center; V A W is an angle of 1¼°, which equals the freedom between the guard point and the roller; *g g* represents the path of the guard pin *u* for the single roller, and is drawn at the intersection of VA with the roller A′ A2 is a line drawn from the balance center through that of the ruby pin, and therefore also passes through the center of the crescent. By planting a compass on this line, where it cuts the periphery of the roller, and locating the point of intersection of VA with the roller, will give us one-half the crescent, the remaining half being transferred to the opposite side of the line A' A2. We will notice that the guard point has entered the crescent 1¼° before the fork begins to move.

The angle of opening for the crescent in the double roller escapement is greater than in the single, because it is placed closer to the balance center, and the guard point or dart further from the pallet center, causing a greater intersection; also the velocity of the guard point has increased, while that of the safety roller has decreased. Fig. 24, at *ff*, shows the path of the dart *h*, which also has 1¼° freedom between bank and roller. From the balance center we draw A' *d* touching the center or point of the dart; from this point we construct at 5° angle *b* A' *d*. This is to ensure sufficient freedom for the dart when entering the crescent. We plant a compass on the point of intersection of A' A2 with the safety roller, S2, and locating the point where A'*b* intersects it, have found one-half the opening for the crescent, the remaining half being constructed on the opposite side of the line A' A2.

THE HORN.

The Horn on the fork belongs to the safety action: more horn is required with the double than with the single roller, on account of the greater angle of opening for the crescent.

The horn should be of such a length that when the crescent has passed the guard point, the end of the horn should point to at least the center of the ruby pin.

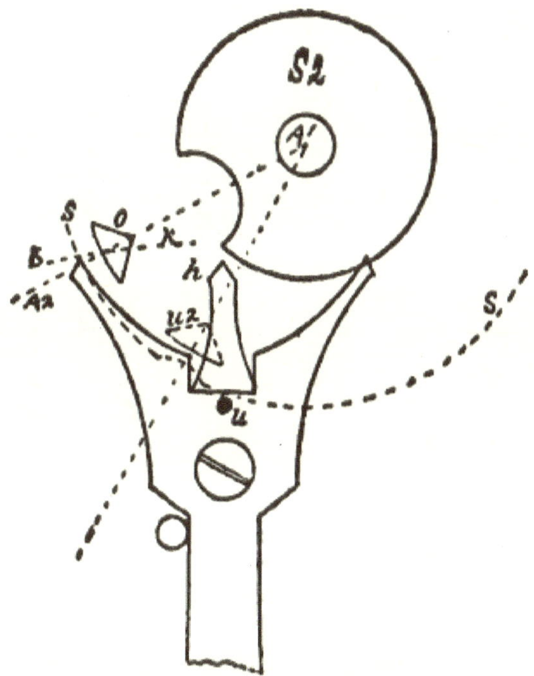

FIG. 25.

The dotted circle, *s s*, Fig. 25, represents a single roller. It will be noticed that the corner of the crescent has passed the guard pin *u* by a considerable angle, and although this is so, in case of an accident the *acting edge* of the fork would come in contact with the ruby pin; this proves that a well made single roller escapement really requires but little horn, only enough to ensure the safe entry of the ruby pin in case the guard point at that moment be thrown against the roller. We will now examine the question from the standpoint of the double roller; S2, Fig. 25, is the safety roller; the corner of the crescent has safely passed the dart *h*; the centers of the ruby pin *o* and of the crescent being on the line A′ A2, we plant the compass on the pallet center and the center of the face of the ruby pin and draw *k k*, which will be the path described by the horn. The end of the horn is therefore planted upon it from 1½° to 1¾° from the ruby pin; this freedom at the end of the horn is therefore from ¼° to ½° more than we allow for the guard point; it depends upon the size of the escapement and locking angles which we would choose. It must in any case be less than the lock on the pallets, so that the fork will be drawn back against the bank in case the horn be thrown against the ruby pin.

When treating on the width of the ruby pin, we mentioned the Savage pin roller escapement, which we illustrate in Figs. 26 and 27. This ingenious arrangement was designed with the view of combining the advantages of both wide and narrow pins and at the same time without any of their disadvantages.

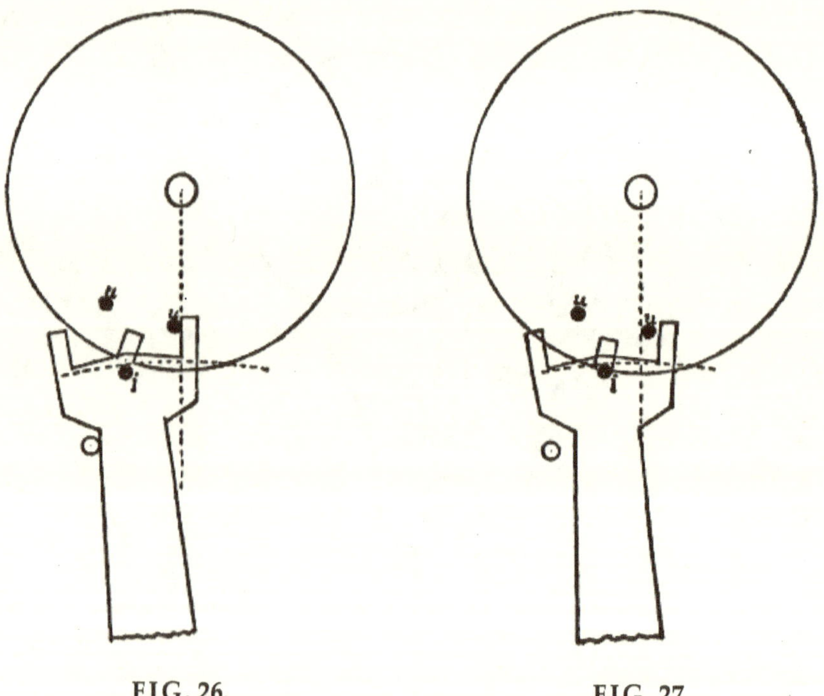

FIG. 26.   FIG. 27.

In Fig. 26 we show the unlocking pins *u* beginning their action on the line of centers—the best possible point—in unlocking the escapement. These pins were made of gold in all which we examined, although it is recorded that wide ruby pins and ruby rollers have been used in this escapement, which would be preferable.

The functions of the two pins in the roller are simply to unlock the escapement; the impulse is not transmitted to them as is the case in the ordinary fork and roller action. In this action the guard pin *i* also acts as the impulse pin. We will notice that the passing hollow in this roller is a rectangular slot the same as in the ordinary fork. When the escapement is being unlocked the guard pin *i* enters the hollow and when the escape tooth comes into contact with the lifting plane of the pallet the pin *i*, Fig. 27, transmits the impulse to the roller.

The impulse is transmitted closer to the line of centers than could be done with any ruby pin. If the pin *i* were wider the impulse would be transmitted still closer to the line of centers, but the intersection of it with the roller would be less. It is very delicate as it is, therefore from a practical standpoint it ought to be made thin but consistent with solidity. If the pin is anyway large, it should be flattened on the sides, otherwise the friction would be similar to that of the round ruby pin. It would also be preferable (on account of the pin *i* being very easily bent) to make the impulse piece narrow but of such a length that it could be screwed to the fork, the same as the dart in the double roller. The impulse radius is also the radius of the roller, because the impulse is transmitted to the roller itself; for this reason the latter is smaller in this action than in the ordinary one having the same angles; also a shorter lever is in contact with a longer one in the unlocking than in ordinary action of the same angles; but for all this the pins *u u* should be pitched close to the edge of the roller, as the angular connection of the balance with the escapement would be increased during the unlocking action. This escapement being very delicate requires a 12° pallet angle and a proportion between impulse and pallet angles of not less than 3 to 1, which would mean an impulse angle of 36°; this, together with the first rate workmanship required are two of the reasons why this action is not often met with.

George Savage, of London, England, invented this action. He was a watchmaker who, in the early part of this century, did much to perfect the lever escapement by good work and nice proportion, besides inventing the two pin variety. He spent the early part of his life in Clerkenwell, but in his old days emigrated to Canada, and founded a flourishing retail business in Montreal, where he died. Some of George Savage's descendants are still engaged at the trade in Canada at the present day.

The correct delineation of the lever escapement is a very important matter. We illustrate one which is so delineated that it can be practically produced. We have not noticed a draft of the lever escapement, especially with equidistant pallets and club teeth, which would act correctly in a watch.

We have been aggressive in our work and have sometimes found theories propounded and elongated which of themselves were not right; this may have something to do with it, that we so often hear workmen say, "Theory is no use, because if you work according to it your machine will not run." We say, "No, sir, if your theory is not right in itself, then your work will certainly not be correct; but if your theory be correct then your work *must* be correct. Why? it simply cannot be otherwise." We will give it another name; let us say, apply sense, reason, thought, experience and study to your work, and what have you done? You have simply applied theory.

A theorem is a proposition to be proved, not being able to prove it, we must simply change it according as our experience dictates, this is precisely what we have done with the escapement after having followed the deductions of recognized authorities with the result that we can now illustrate an escapement which has been thoroughly subjected to an impartial analysis in every respect, and which is theoretically and practically correct.

We will not only give instructions for drafting the escapement now under consideration, but will also make explanations how to draft it in different positions, also in circular pallet and single roller. We are convinced that by so doing we will do a service to many, we also wish to avoid what we may call "the stereotyped" process, that is, one which may be acquired by heart, but introduce any changes and perplexity is the result. It is really not a difficult matter to draft escapements in different positions, as an example will show.

Before making a draft we must know exactly what we wish to produce. It is well in drafting escapements to make them as large as possible, say thirty to forty times larger than in the watch, in the present case the size is immaterial, but we must have specifications for the proportions of the angles. Our draft is to be the most difficult subject in lever escapements; it is to be represented just as if it were working in a watch; it is to represent a good and reliable action in every respect, one which can be applied without special difficulty to a good watch, and is to be "up to date" in every particular and to contain the majority of the best points and conclusions reached in our analysis.

SPECIFICATIONS FOR LEVER ESCAPEMENT.

Specifications for Lever Escapement: The pallets are to be equidistant; the wheel teeth of the "club" form; there are to be two rollers; wheel, pallet, and balance centers are to be in straight line. The lock is to be 1½°, the run ¼°, making a total lock of 1¾°; the movement of pallets from drop to drop is to be 10°, while the fork is to move through 10¼° from bank to bank; the lift on the wheel teeth is to be 3°, while the remainder is to be the lift on the pallets as follows: 10¼ − (1¾ + 3) = 5½° for lift of pallets.

The wheel is to have 15 teeth, with pallets spanning 3 teeth or 2½ spaces, making the angle from lock to lock = 360 ÷ 15 × 2½ = 60°, the interval from tooth to tooth is 360 ÷ 15 = 24°; divided by 2 pallets = 24 ÷ 2 = 12° for width of tooth, pallet and drop; drop is to be 1½°, the tooth is to be ¾ the width of the pallet, making a tooth of a width of 4½° and a pallet of 6°.

The draw is to be 12° on each pallet, while the locking faces of the teeth are to incline 24°. The acting length of fork is to be equal to the distance of centers of scape wheel and pallets; the impulse angle is to be 28°; freedom from dart and safety, roller is to be 1¼°, and for dart and corner of crescent 5°; freedom for ruby pin and acting edge of fork is to be 1¼°; width of slot is to be ½ the total motion, or 10¼ ÷ 2 = 5⅛°; shake of ruby pin in slot = ¼°, leaving 5⅛ − ¼ = 4⅞° for width of ruby pin.

Radius of safety roller to be $\frac{4}{7}$ of the theoretical impulse radius. The length of horn is to be such that the end would point at least to the center of the ruby pin when the edge of the crescent passes the dart; space between the end of horn and ruby pin is to be 1½°.

It is well to know that the angles for width of teeth, pallets and drop are measured from the wheel center, while the lifting and locking angles are struck from the pallet center, the draw from the locking corners of the pallets, and the inclination of the teeth from the locking edge.

In the fork and roller action, the angle of motion, the width of slot, the ruby pin and its shake, the freedom between dart and roller, of ruby pin with acting edge of fork and end of horn are all measured from the pallet center, while the impulse angle and the crescent are measured from the balance center. A sensible drawing board measures 17 × 24 inches, we also require a set of good drawing instruments, the finer the instruments the better; pay special attention to the compasses, pens and protractor; add to this a straight ruler and set square.

The best all-round drawing paper, both for India ink and colored work has a rough surface; it must be fastened firmly and evenly to the board by means of thumb tacks; the lines must be light and made with a hard pencil. Use Higgins' India ink, which dries rapidly.

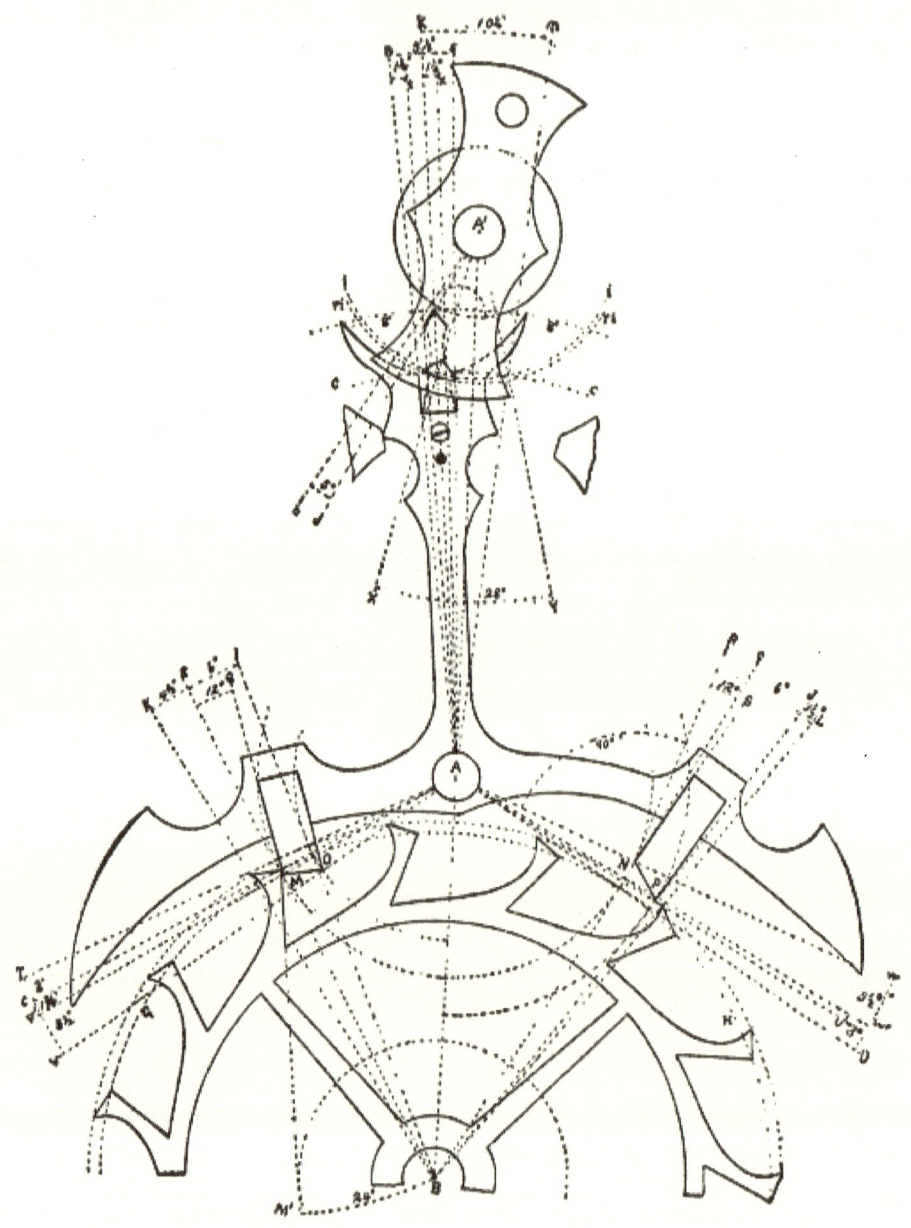

We will begin by drawing the center line A' A B; use the point B for the escape center; place the compass on it and strike G H, the primitive or geometrical circle of the escape wheel; set the center of the protractor at B and mark off an angle of 30° on each side of the line of centers; this will give us the angles A B E and A B F together, forming the angle F B E of 60°, which represents from lock to lock of the pallets. Since the chord of the angle of 60° is equal to the radius of the circle, this gives us an easy means of verifying this angle by placing the compass at the points of intersection of F B and E B with the primitive circle G H; this distance must be equal to the radius of the circle. At these points we will construct right angles to E B and F B, thus forming the tangents C A and D A to the primitive circle G H. These tangents meet on the line of centers at A, which will be the pallet center. Place the compass at A and draw the locking circle M N at the points of intersection of E B and F B with the primitive circle G H. The locking edges of the pallets will always stand on this circle no matter in what relation the pallets stand to the wheel. Place the center of the protractor at B and draw the angle of width of pallets of 6°; I B E being for the engaging and J B F for the disengaging pallet. In the equidistant pallet I B is drawn on the side towards the center, while J B is drawn further from the center. If we were drawing a circular pallet, one-half the width of pallets would be placed on each side of E B and F B. At the points of intersection of I B and J B with the primitive circle G H we draw the path O for the discharging edge of the engaging and P for that of the disengaging pallet. The total lock being 1¾°, we construct V′ A at this angle from C A; the point of intersection of V′ A with the locking circle M N, is the position of the locking corner of the engaging pallet. The pallet having 12° draw when locked we place the center of the protractor on this corner and draw the angle Q M E. Q M will be the locking face of the engaging pallet. If the face of the pallet were on the line E B there would be no draw, and if placed to the opposite side of E B the tooth would repel the pallet, forming what is known as the repellant escapement.

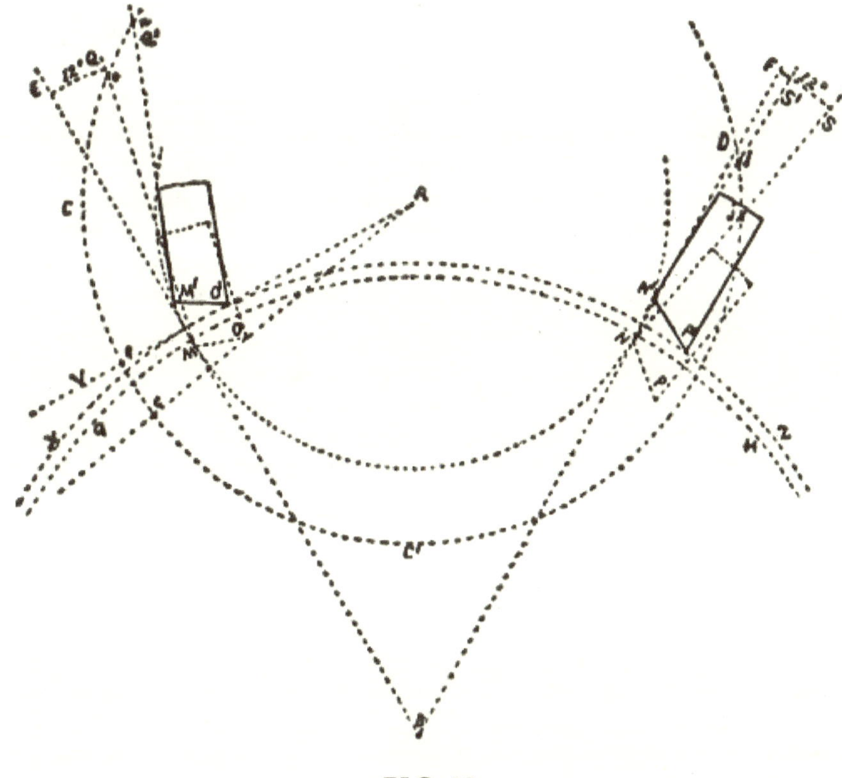

FIG. 28.

Having shown how to delineate the locking face of the engaging pallet when locked, we will now consider how to draft both it and the disengaging pallet in correct positions when unlocked; to do so we direct our attention until further notice to Fig. 28. The locking faces Q M of the engaging and S N of the disengaging pallets are shown in dotted lines *when locked*. We must now consider the relation which the locking faces will bear to E B in the engaging, and to F B in the disengaging pallets when unlocked. This is a question of some importance; it is easy enough to represent the 12° from the 30° angles when locked; we must be certain that they would occupy exactly that position and yet show them unlocked; we shall take pains to do so. In due time we shall show that there is no appreciable loss of lift on the engaging pallet in the escapement illustrated; the angle T A V therefore shows the total lift; we have not shown the corresponding angles on the disengaging side because the angles are somewhat different, but the total lift is still the same. G H represents the primitive circle of the escape wheel, and X Z that of the real, while M N represents the circular course which the locking corners of the pallets take in an equidistant escapement. At a convenient position we will construct the circle C C' D from the pallet center A. Notice the points *e* and *c*, where V A and T A intersect this circle; the space between *e* and *c* represents the extent of the motion of the pallets at this particular distance from the center A; this being so, then let us apply it to the engaging pallet. At the point of intersection *o* of the dotted line Q M (which is an extended line on which the face of the pallet lies when locked), with the circle C C' D, we will plant our dividers and transfer *e c* to *o n*. By setting our dividers on *o* M and transferring to *n* M', we will obtain the location of Q' M', the locking face when unlocked. Let us now turn our attention to the disengaging pallet. The dotted line S N represents the location of the locking face of the disengaging pallet when locked at an angle of 12° from F B. At the intersection of S N with the circle C C' D we obtain the point *j*. The motion of the two pallets being equal, we transfer the distance *e c* with the dividers from *j* and obtain the point *l*. By setting the dividers on *j* N and transferring to *l* N' we draw the line S' N' on which the locking face of the disengaging pallet will be located when unlocked. It will be perfectly clear to anyone that through these means we can correctly represent the pallets in any desired position.

We will notice that the face Q' M' of the engaging pallet when unlocked stands at a greater angle to E B than it did when locked, while the opposite is the case on the disengaging pallet, in which the angle S' N' F is much less than S N F. This shows that the *deeper* the engaging pallet locks, the lighter will the draw be, while the opposite holds good with the disengaging pallet; also, that the draw increases during the unlocking of the engaging, and decreases during the unlocking of the disengaging pallet. These points show that the draw should be measured with the *fork standing against the bank*; not when the locking corner of the pallet stands on the primitive circle, as is so often done. The recoil of the wheel (which determines the draw), is illustrated by the difference between the locking circle M N and the face Q M for the engaging, and S N for the disengaging pallet, and along the *acting* surface it is alike on each pallet, showing that the draft angle should be the same on each pallet.

A number of years ago we constructed the escapement model which we herewith illustrate. All the parts are adjustable; the pallets can be moved in any direction, the draft angles can be changed at will. Through this model we can practically demonstrate the points of which we have spoken. Such a model can be made by workmen after studying these papers.

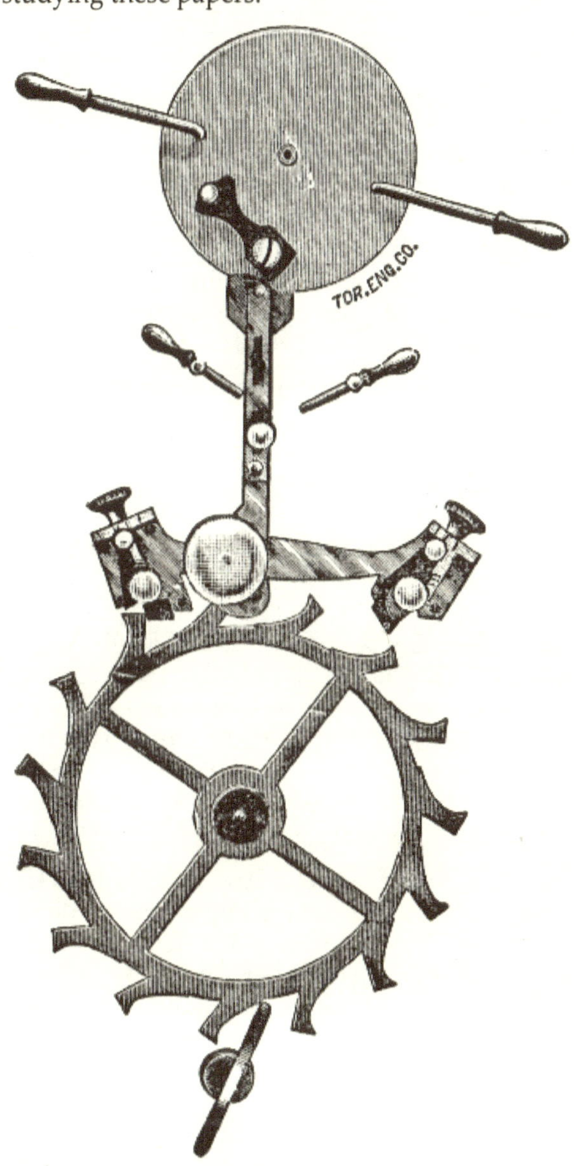

In both the equidistant and circular pallets the locking face S N of the disengaging pallet deviates more from the locking circle M N than does the locking face Q M of the engaging pallet, as will be seen in the diagram. This is because the draft angle is struck from E B which deviates from the locking circle in such a manner, that if the face of a pallet were planted on it and *locked deep enough* to show it, the wheel would actually *repel* the pallet, whereas with the disengaging pallet if it were planted on F B, it would actually produce draw if locked very deep; this is on account of the natural deviation of the 30° lines from the locking circle. This difference is more pronounced in the circular than in the equidistant pallet, because in the former we have two locking circles, the larger one being for the engaging pallet, and as an arc of a large circle does not deviate as much from a straight line as does that of a smaller circle, it will be easily understood that the natural difference before spoken of is only enhanced thereby. For this reason in order to produce an *actual* draw of 12°, the engaging pallet may be set at a slightly greater angle from E B in the circular escapement; the amount depends upon the width of the pallets; the requirements are that the recoil of the wheel will be the same on each pallet. We must, however, repeat that one of the most important points is to measure the draw when the fork stands against the bank, thereby *increasing* the draw on the engaging and *decreasing* that of the disengaging pallet *during* the unlocking action, thus *naturally* balancing one fault with another.

We will again proceed with the delineation of the escapement here illustrated. After having drawn the locking face Q M, we draw the angle of width of teeth of 4½°, by planting the protractor on the escape center B. We measure the angle E B K, from the locking face of the pallet; the line E B does not touch the locking face of the pallet at the present time of contact with the tooth, therefore a line must be drawn from the point of contact to the center B. We did so in our drawing but do not illustrate it, as in a reduced engraving of this kind it would be too close to E B and would only cause confusion. We will now draw in the lifting angle of 3° for the tooth. From the tangent C A we draw T A at the required angle; at the point of intersection of T A with the 30° line E B we have the real circumference of the escape wheel. It will only be necessary to connect the locking edge of the tooth with the line K B, where the real or outer circle intersects it. It must be drawn in the same manner in the circular escapement; if the tooth were drawn up to the intersection of K B with T A, the lift would be too great, as that point is further from the center A than the points of contact are.

If the real or outer circle of the wheel intersects both the locking circle M N and the path O of the discharging edge at the points where T A intersects them, then there will be *no loss* of lift on the engaging pallet. This is precisely how it is in the diagram; but if there is any deviation, then the angle of loss must be measured on the *real* diameter of the wheel and not on the primitive, as is usually done, as the real diameter of the wheel, or in other words the heel of the tooth, forms the last point of contact. With a wider tooth and a greater lifting angle there will even be a *gain* of lift on the engaging pallet; the pallet in such a case would actually require a smaller lifting angle, according to the amount of gain. We gave full directions for measuring the loss when describing its effects in Fig. 8. Whatever the loss amounts to, it is added to the lifting plane of the pallet. In the diagram under discussion there is no loss, consequently the lifting angle on the pallet is to be 5½°. From V' A we draw V A at the required angle; the point of intersection of V A with the path O will be the discharging edge O. It will now only be necessary to connect the locking corner M with it, and we have the lifting plane of the pallet; the discharging side of the pallet is then drawn parallel to the locking face and made a suitable length. We will now draw the locking edges of the tooth by placing the center of the protractor on the locking edge M and construct the angle B M M' of 24° and draw a circle from the scape center B, to which the line M M' will be a tangent. We will utilize this circle in drawing in the faces of the other teeth after having spaced them off 24° apart, by simply putting a ruler on the locking edges and on the periphery of the circle.

We now construct W' A as a tangent to the outer circle of the wheel, thus forming the lifting angle D A W' of 3° for the teeth; this corresponds to the angle T A C on the engaging side. W' A touches the outer circle of the wheel at the intersection of F B with it. We will notice that there is considerable deviation of W' A from the circle at the intersection of J B with it. At the intersecting of this point we draw U A; the angle U A W' is the loss of lift. This angle must be added to the lifting angle of the pallets; we see that in this action there is no loss on the engaging pallet, but on the disengaging the loss amounts to approximately ⅞° in the action illustrated. As we have allowed ¼° of run for the pallets, the discharging edge P is removed at this angle from U A; we do not illustrate it, as the lines would cause confusion being so close together. The lifting angle on the pallet is measured from the point P and amounts to 5½° + the angle of the loss; the angle W A U embraces the above angles besides ¼° for run. If the locks are equal on each pallet, it proves that the lifts are also equal. This gives us a practical method of proving the correctness of the drawing; to do so, place the dividers on the locking circle M N at the intersection of T A and V A with it, as this is the extent of motion; transfer this measurement to N, if the *actual* lift is the same on each pallet, the dividers will locate the point which the locking corner N will occupy *when locked*; this, in the present case, will be at an angle of 1¾° below the tangent D A. By this simple method, the correctness of our proposition that the loss of lift should be measured from the outside circle of the wheel, can be proven. We often see the loss measured for the engaging pallet on the primitive circumference G H, and on the real circumference for the disengaging; if one is right then the other must be wrong, as there is a noticeable deviation of the tangent C A from the primitive circle G H at the intersection of the locking circle M N; had we added this amount to the lifting angle V' A V of the engaging pallet, the result would have been that the discharging edge O would be over 1° below its present location, thus showing that by the time the lift on the engaging pallet had been completed, the locking corner N of the disengaging pallet would be locked at an angle of 2¾° instead of only 1¾°. Many watches contain precisely this fault. If we wish to make a draft showing the pallets at any desired position, at the center of motion for instance, with the fork standing on the line of centers, we would proceed in the following manner: 10¼° being the total motion, one-half would equal 5⅛°; as the total lock equals 1¾°, we deduct this amount from it which leaves 5⅛ − 1¾ = 3⅜°, which is the angle at which the locking corner M should be shown above the tangent C A. Now let us see where the locking corner N should stand; M having moved up 5⅛°, therefore N moved down by that amount, the lift on the pallet being 5½° and on the tooth 3° (which is added to the tangent D A), it follows that N should stand 5½ + 3 − 5⅛ = 3⅜° above D A. We can prove it by the lock, namely: 3⅜° + 1¾ = 5⅛°, half the remaining motion. This shows how simple it is to draft pallets in various positions, remembering always to use the tangents to the primitive circle as measuring points. We have fully explained how to draw in the draft angle on the pallets when unlocked, and do not require to repeat it, except to say, that most authorities draw a tangent R N to the locking circle M N, forming in other words, the right angle R N A, then construct an angle of 12° from R N. We have drawn ours in by our own method, which is the correct one. While we here illustrate S N R at an angle of 12° it is in reality *less* than that amount; had we constructed S N at an angle of 12° from R N, then the draw would be 12° from F B, when the primitive circumference of the wheel is reached, but *more* than 12° when the fork is against the bank.

The space between the discharging edge P and the heel of the tooth forms the angle of drop J B I of 1½°; the definition for drop is that it is the freedom for wheel and pallet. This is not, strictly speaking, perfectly correct, as, during the unlocking action there will be a recoil of the wheel to the extent of the draft angle; the heel of the tooth will therefore approach the edge P, and the discharging side of the pallet approaches the tooth, as only the discharging edge moves on the path P.

A good length for the teeth is $\frac{1}{10}$ the diameter of the wheel, measured from the primitive diameter and from the locking edge of the tooth.

The backs of the teeth are hollowed out so as not to interfere with the pallets, and are given a nice form; likewise the rim and arms are drawn in as light and as neat as possible, consistent with strength.

Having explained the delineation of the wheel and pallet action we will now turn our attention to that of the fork and roller. We tried to explain these actions in such a manner that by the time we came to delineate them no difficulty would be found, as in our analysis we discussed the subject sufficiently to enable any one of ordinary intelligence to obtain a correct knowledge of them. The fork and roller action in straight line, right, or any other angle is delineated after the methods we are about to give.

We specified that the acting length of fork was to be equal to the center distance of wheel and pallets; this gives a fork of a fair length.

Having drawn the line of centers A' A we will construct an angle equal to half the angular motion of the pallets; the latter in the case under consideration being 10¼°, therefore 5⅛° is spaced off on each side of the line of centers, forming the angles *m* A *k* of 10¼°. Placing our dividers on A B the center distance of 'scape wheel and pallets, we plant them on A and construct *c c*; thus we will have the acting length of fork and its path. We saw in our analysis that the impulse angle should be as small as possible. We will use one of 28° in our draft of the double roller; we might however remark that this angle should vary with the construction of the escapements in different watches; if too small, the balance may be stopped when the escapement is locked, while if too great it can be stopped during the lift; both these defects are to be avoided. The angles being respectively 10¼° and 28° it follows they are of the following proportions: 28° ÷ 10.25 = 2.7316. The impulse radius therefore bears this relation (but in the inverse ratio to the angles), to the acting length of fork.

We will put it in the following proportion; let A c equal acting length of fork, and *x* the unknown quantity; 28:10.25 :: A c:x; the answer will be the theoretical impulse radius. Having found the required radius we plant one jaw of our measuring instrument on the point of intersection of *c c* with *k* A or *m* A and locate the other jaw on the line of centers; we thus obtain A' the balance center. Through the points of intersection before designated we will draft X A' and Y A' forming the impulse angle X A' Y of 28°. At the intersection of this angle with the fork angle *k* A' *m*, we draw *i i* from the center A; this gives us the theoretical impulse circle. The total lock being 1¾° it follows that the angle described by the balance in unlocking = 1¾ × 2.7316 = 4.788°. According to the specifications the width of slot is to be 5⅛°; placing the center of the protractor on A we construct half of this angle on each side of *k* A, which passes through the center of the fork when it rests against the bank; this gives us the angle *s* A *n* of 5⅛°. If the disengaging pallet were shown locked then *m* A would represent the center of the fork. The slot is to be made of sufficient depth so there will be no possibility of the ruby pin touching the bottom of it. The ruby pin is to have 1¼° freedom in passing the acting edge of the fork; from the center A we construct the angle *t* A *n* of 1¼°; at the point of intersection of *t* A with *c c* the acting radius of the fork, we locate the real impulse radius and draw the arc *ri ri* which describes the path made by the face of the ruby pin. The ruby pin is to have ¼° of shake in the slot; it will therefore have a width of 4⅞°; this width is drawn in with the ruby pin imagined as standing over the line of centers and is then transferred to the position which the ruby pin is to occupy in the drawing.

The radius of the safety roller was given as ⁴∕₇ of the theoretical impulse radius. They may be made of various proportions; thus ⅔ is often used. Remember that the smaller we make it, the less the friction during accidental contact with the guard pin, the greater must the passing hollow be and the horn of fork and guard point must be longer, which increases the weight of the fork.

Having drawn in the safety roller, and having specified that the freedom between the dart and safety roller was to be 1¼°, the dart being in the center of the fork, consequently *k* A is the center of it; therefore we construct the angle *k* A X of 1¼°. At the point of intersection of X A with the safety roller we draw the arc *g g*; this locates the point of the dart which we will now draw in. We will next draw *d* A' from the balance center and touching the point of the dart; we now construct *b* A' at an angle of 5° to it. This is to allow the necessary freedom for the dart when entering the crescent; from A' we draw a line through the center of the ruby pin. We do not show it in the drawing, as it would be indiscernible, coming very close to A' X. This line will also pass through the center of the crescent. At the point of intersection of A' *b* with the safety roller we have one of the edges of the crescent. By placing our compass at the center of the crescent on the periphery of the roller and on the edge which we have just found, it follows that our compass will span the radius of the crescent. We now sweep the arc for the latter, thus also drawing in the remaining half of the crescent on the other side of A' X and bringing the crescent of sufficient depth that no possibility exists of the dart touching in or on the edges of it. We will now draw in the impulse roller and make it as light as possible consistent with strength. A hole is shown through the impulse roller to counterbalance the reduced weight at the crescent. When describing Fig. 24, we gave instructions for finding the dimensions of crescent and position of guard pin for the single roller. We will find the length of horn; to do so we must closely follow directions given for Fig. 25. In locating the end of the horn, we must find the location of the center of the crescent and ruby pin *after* the edge of the crescent has passed the dart. From the point of intersection of A′ b with the safety roller we transfer the radius of the crescent on the periphery of the safety roller towards the side against the bank, then draw a line from A′ through the point so found. At point of intersection of this line with the real impulse circle r i r i we draw an arc radiating from the pallet center; the end of the horn will be located on this arc. In our drawing the arc spoken of coincides with the dart radius g g. As before pointed out, we gave particulars when treating on Fig. 25, therefore considered it unnecessary to further complicate the draft by the addition of all the constructional lines. We specified that the freedom between ruby pin and end of horn was to be 1½°; these lines (which we do not show) are drawn from the pallet center. Having located the end of the horn on the side standing against the bank, we place the dividers on it and on the point of intersection of k A with g g—which in this case is on the point of the dart,—and transfer this measurement along g g which will locate the end of the horn on the opposite side.

We have the acting edges of the fork on cc and have also found the position of the ends of the horns; their curvature is drawn in the following manner: We place our compasses on A and r i, spanning therefore the real impulse radius; the compass is now set on the acting edge of the fork and an arc swept with it which is then to be intersected by another arc swept from the end of the horn, on the same side of the fork. At the point of intersection of the arcs the compass is planted and the curvature of the horn drawn in, the same operation is to be repeated with the other horn. We will now draw in the sides of the horn of such a form that should the watch rebank, the side of the ruby pin will squarely strike the fork. If the back of the ruby pin strikes the fork there will be a greater tendency of breaking it and injuring the pivots on account of acting like a wedge. The fork and pallets are now drawn in as lightly as possible and of such form as to admit of their being readily poised. The banks are to be drawn at equal distances from the line of centers. In delineating the fork and roller action in any desired position, it must be remembered that the points of location of the real impulse radius, the end of horn, the dart or guard pin and crescent, must *all* be obtained *when standing against the bank,* and the arcs drawn which they describe; the parts are then located according to the angle at which they are removed from the banks.

We think the instructions given are ample to enable any one to master the subject. We may add that when one becomes well acquainted with the escapement, many of the angles radiating from a common center, may be drawn in at once. We had intended describing the mechanical construction of the escapement, which does unmistakably present some difficulties on account of the small dimensions of the parts, but nevertheless it can be mechanically executed true to the principles enumerated. We have evolved a method of so producing them that young men in a comparatively short period have made them from their drafts (without automatic machinery) that their watches start off when run down the moment the crown is touched. Perhaps later on we will write up the subject. It is our intention of doing so, as we make use of such explanations in our regular work.

Lector House believes that a society develops through a two-fold approach of continuous learning and adaptation, which is derived from the study of classic literary works spread across the historic timeline of literature records. Therefore, we aim at reviving, repairing and redeveloping all those inaccessible or damaged but historically as well as culturally important literature across subjects so that the future generations may have an opportunity to study and learn from past works to embark upon a journey of creating a better future.

This book is a result of an effort made by Lector House towards making a contribution to the preservation and repair of original ancient works which might hold historical significance to the approach of continuous learning across subjects.

HAPPY READING & LEARNING!

LECTOR HOUSE LLP
E-MAIL: lectorpublishing@gmail.com